# VITAMIN A SKIN SCIENCE

Authors: Dr Desmond Fernandes and Dr Ernst Eiselen

# Table of CONTENTS

| CHAPTER | | PAGE |
|---|---|---|
| | A First Word by Dr Ernst Eiselen | 08 |
| 01 | Introduction and Background | 10 |
| 02 | A Perspective on Modern Skin Care | 16 |
| 03 | The Structure of the Skin | 21 |
| 04 | The Sun and Free Radicals | 36 |
| 05 | Photoageing | 45 |

# VITAMIN A SKIN SCIENCE

Authors: Dr Desmond Fernandes and Dr Ernst Eiselen

Copyright © 2015
Fernro Publishing Ltd.

All rights reserved. No reproduction, copy or transmission of this publication may be made without written permission. No paragraph of this publication may be reproduced, copied, or transmitted save with the written permission or in accordance with the provisions of the Copyright, Designs and Patents Act 1988, or under the terms of any licence permitting limited copying issued by the Copyright Licensing Agency.

The authors and publishers have made every effort to ensure the accuracy of the information herein. However, the information is sold without warranty, either express or implied. Neither the authors, Fernro Publishing Ltd. nor its distributors will be held liable for any damages to be caused either directly or indirectly by any instructions or information contained in this book.

Any person who does any unauthorised act in relation to this publication may be liable to criminal prosecution and civil claims for damages. The authors have asserted their rights to be identified as the authors of this work in accordance with the Copyright, Designs and Patents Act 1988.

First published May 2015
Fernro Publishing Ltd.
London
United Kingdom

ISBN 978-0-9576681-9-5

This book is sold subject to the condition that it shall not, by way of trade or otherwise, be lent, re-sold, hired out, or otherwise circulated without the publisher's prior consent in any form of binding or cover other than that in which it is published, and without a similar condition including this condition being imposed on the subsequent purchaser.

| CHAPTER | | PAGE |
|---|---|---|
| 06 | The Prevention of Photoageing: Sun Protection | 54 |
| 07 | The Chemistry of Vitamin A | 64 |
| 08 | The Role of Vitamin A in Photoageing | 79 |
| 09 | My Personal Experience with Using Vitamin-A-Based Skin Care Products, by Dr Des Fernandes | 89 |
| 10 | The Safety of Vitamin A by Mouth and Applied to the Skin | 100 |
| 11 | Antioxidant Vitamins: C, E, and Others | 106 |
| 12 | Other Important Molecules for Protecting Skin | 122 |
| 13 | Peptides | 127 |
| 14 | Other Molecules to Rejuvenate Skin: Alpha- and Beta-Hydroxy Acids and Others | 138 |
| 15 | Skin Peeling | 150 |
| 16 | When Skin Care is not Enough: Enhanced Skin Penetration with Iontophoresis and Sonophoresis | 157 |
| 17 | The Principles of Cellular Communication | 181 |
| 18 | Skin Needling | 188 |
| 19 | A Final Word, by Dr Des Fernandes | 204 |
| Index | | 206 |

# A First Word by
# DR ERNST EISELEN

Dr Des Fernandes set out in the 1980s to find a way to prevent sun-damaged skin from making malignant melanomas. He had no idea that this simple yet highly ethical and compassionate desire would start a journey ultimately leading him to ground-breaking skin treatments.

Dr Fernandes' unravelling of skin cell behaviour and relationships, which depend largely on skin micro-nutrition as described in this book, is the key to successful treatment. These treatments determine whether one will reside in a resilient, comfortable, and beautiful skin or skin that prematurely succumbs to the ravages of age and environment.

Although this may sound somewhat far-fetched, the solid base of published dermatological, medical, and other scientific research over the last six decades fully supports the philosophy and practice of vitamin A as the kingpin in cell differentiation. The paramount role of vitamin A in human existence and the pivotal roles of vitamin C and vitamin E in the skin and other human tissue are no longer open to question. No other skin treatment technology bases its formulations and treatment interventions as purely on the genetically inherited primary physiology of skin cells and wound repair, as those developed by Dr Fernandes. Furthermore, these processes and systems are genetically inherited by human generations and are thus unalterable.

This book is unique in the insights it provides to make sense of skin care at a time when people with vulnerable skins grow much older than in the past and are exposed to many more hours of damaging sunlight and artificial light. The reader is provided with logical explanations and an understanding of how the skin functions on a fundamental level. Knowledge like this provides a path of understanding leading one out of the quagmire of clever marketing and high-pressure advertising. As the distinction between highly active cosmetic products and medications progressively blurs, a guide to understanding the necessary process of skin health comes not before time.

A unique source of information, the text provides a veritable 'manual' for anyone wishing to improve skin to a superior quality at any age. Plain scientific logic and evidence-based skin physiology untainted by wild claims and 'single ingredient worship', lays bare indispensable facts about the largest organ we humans possess. The book makes clear how to live in healthy skin, how to address damage, and how to take steps to improve skin for the future.

This book is an eloquent exposition of a momentous body of work in progress. It is the fruit of more than two decades of honest, forthright, and innovative research into the most practical and affordable ways to achieve and maintain a life-long comfortable, beautiful, and healthy skin. There is no 'fluff', no semantics of selling – just plain science and logical thinking.

*"The paramount role of vitamin A in human existence and the pivotal roles of vitamin C and E in the skin and other human tissue are no longer open to question."*

Dr. Ernst Eiselen
June 2014.
Perth WA

# Chapter 01
# INTRODUCTION AND BACKGROUND

To many people it may seem strange that two South African doctors should write a book about scientific skin care for the Internet Age. They may also question why anyone from a 'third world' country should be so presumptuous as to believe that the science of skin care devised in the very southern tip of Africa could be the best for people living in the most advanced parts of the world. The truth is that the fundamental physiology of skin applies to all humans, no matter what colour their skin or what climatic conditions they live under.

Here is a short story from each of us to tell you who we are and what we do and what we believe in when it comes to skin care. Des is a plastic surgeon practising in Cape Town, South Africa, and Ernst is a specialist general practitioner practising in Perth, Australia, previously from Kwa-Zulu Natal, South Africa.

## DES

In 1981 I started to take an interest in skin particularly because it became apparent that malignant melanoma was becoming more prevalent than it had been in the 1960s. I went regularly to the medical library of the University of Cape Town to read about sun damage because of a very sad series of events in 1979 that catapulted me into researching malignant melanoma. Two young patients changed my world. They came in for routine removal of pigmented naevi (moles), but these very sadly turned out to be aggressive malignant melanomas.

One was a beautiful 19-year-old girl who had developed a pigmented mark on her arm while she was an exchange student in the USA. The other person was a handsome 20-year-old male student who ignored a pigmented mole that had grown over a few months. Both these young people were intelligent, dynamic and charismatic and we fought hard to cure their melanomas. They were dead before either had reached the age of 23.

The sense of failure hung like a heavy cloud over me, and I felt I needed to understand more about malignant melanoma and the sun damage that caused these dreadful cancers. As we had also recently learned about the hole in the ozone layer over the Antarctic, it was natural to assume that atmospheric changes were part of the cause. My reading of thick reference books led me to the unfolding of an understanding of the miraculous effects of vitamin A.

In those days, although I was using vitamin A *acid*, I knew very little about vitamin A. In fact, I knew only what plastic surgeons and dermatologists generally knew – which turned out to be virtually nothing. Vitamin A was also considered simply as a vitamin, and we thought that we knew all we needed to know about it. In addition, I came across a new concept, which is that free radicals cause diseases. When I told my colleagues about free radicals they used to mock me because they 'knew' that bacteria, fungi, or viruses caused diseases. They would not accept that mere electrons could possibly cause disease. Fortunately their attitude simply encouraged me to find out more and more.

All of my research led me to discover a vast amount of information about skin and photodamage that was either unknown to the makers of cosmetic creams, or being deliberately ignored by them. After a few years of intense research into the subject, I felt that I could formulate scientific skin care that would help minimize and maybe even reverse photodamage. I did not particularly want to make the cream myself, and instead felt that cosmetic manufacturers should know about my research. I naïvely believed they would jump at the chance to get their hands on a formula that would produce truly effective skin care products. I wrote to two companies, giving them the principles of my concept, but they replied that they were not interested. I never expected to enter the field of manufacturing cosmetic creams, yet I believed that I should formulate some creams for use on my own patients.

At first I made the creams in my kitchen using rather primitive methods. Understandably, my patients were initially not very enthusiastic about the creams, yet when they started to see the results they became very excited. I soon realized that I needed a cosmetic chemist who could make such sophisticated creams. After I found a chemist the creams were indeed being made by a professional, but still on a small scale. I

*"My reading of thick reference books led me to the unfolding of an understanding of the miraculous effects of vitamin A."*

*This is typical photo-damaged skin seen in type II Fitzpatrick skin in sunny climates. Six months use of vitamin A, C, E and B5 and beta carotene morning (with only SPF 4) and evening made the difference.*

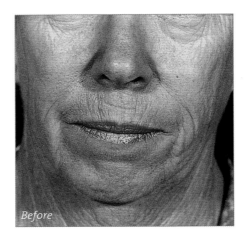
*Before*

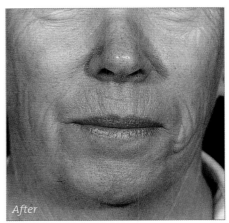
*After*

gave these creams away to patients to use as part of the service I offered for professional advice about rejuvenative facial surgery. They only paid for the creams if they wanted to continue using them after their surgery was completed.

I should mention that I had a very busy and successful plastic surgery practice, and in those days I had a waiting list of about four months for consultations. The waiting list for surgery was up to six months! I found that more and more people were making appointments simply to get hold of the cosmetic creams. They were not at all interested in having surgery. The great results of the creams were so obvious that my patients' friends were desperate to get their hands on them, too. This development was definitely not good for my practice, as the waiting list grew ever longer. Then it dawned on me that the only solution was to make these creams commercially available.

The infant product grew effortlessly without any advertising, simply because of the magnificent results so apparent to everyone who used it. Word of mouth, after all, is a very powerful force. The unusually good skin changes occurred because I had employed good science and a profound understanding of photodamage in the formulation of those very early products.

By 1989 I had been to Japan about eight times and had given creams to my friend Dr Tozawa in Tokyo. She surprised and delighted me by discovering how effective these formulations were for the treatment of skin atopy. 'Atopy' means that skin is sensitized to react to allergens and will usually develop some form of dermatitis or inflammation with a poor quality protective stratum corneum. It seemed entirely surprising that a cream designed for minimizing photodamage also minimized atopy. However, by understanding the actions of vitamin A and the associated antioxidant vitamins, I quickly realized why these changes occurred. I also realized that Japanese skin requires special attention, which

Before

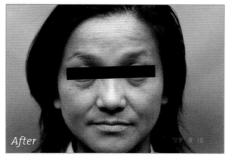
After

*Before: Atopy treated for 40 years with cortisone. After 11 months use of vitamin A, C, E, B5 and beta carotene morning and evening. SPF 4 was used every morning*

meant different formulae, specifically designed for it. Naturally, the Japanese experience reinforced the experience I had in Cape Town, and convinced me that I was ultimately on the right track.

### ERNST

I have been in GP practice all my life, dealing with the enormous spectrum of medical problems that is thrown at a South African GP in a semi-rural setting. I have always had to deal with dermatological problems where I practised, and in particular with skin cancer and sun-damaged skin.

Repeatedly freezing multiple pre-malignant skin lesions was an important, but frustrating, part of my work. Frustrating, as it felt that while it was worthwhile to prevent the skin cancers from forming by freezing lesions with liquid nitrogen, the treatments did nothing to improve or restore the general quality and resilience of the damaged skin as a whole. After I had heard about Dr Fernandes' work in this field, I managed to attend a lecture he gave at Westville Hospital in Durban. I knew there and then that I had found the answer I was looking for.

Patients immediately benefited from using the topical vitamins and a whole new chapter opened in my professional life. In 2002 I attended the Environ international conference in Windhoek in Namibia. Some time afterwards I expressed an interest in being involved with the company as an ad hoc medical advisor on skin matters.

Cell signalling in skin remains my deep interest, temporarily interrupted by the business of moving to Australia in 2007 and getting settled here. My conviction is that there is no need for doctors or scientists to harvest stem cells from anywhere when attempting to repair tissue. All of the information remains preserved in all of our cells. After all, each cell we have stems from the initial, single fertilised ovum that started our lives.

We simply need to discover and understand the details of the 'cell language' to instruct cells to read the appropriate 'chapters' in their DNA. This will alter adult or aged cells to behave in a juvenile fashion. This is a process currently called 'de-differentiation'.

The formulations used in certain specific topically applied vitamin

> *"We simply need to discover and understand the details of the 'cell language' to instruct cells to read the appropriate chapters in their DNA."*

> "New discoveries merely change our ability to understand the intricate chemical pathways that are at work in the cells of the skin."

formulations are an important first step in skin care to achieve this goal. The novelty here is not primarily just the specific molecules used, as these have been part of the lives of our skin cells since the first humans, but their specific combinations and concentrations.

Dr Fernandes has made important practical strides to create skin care that incorporates the fundamentals of this philosophy. I am most excited about what lies ahead of us.

## DES

Formulating truly scientific skin care has been an exciting challenge over the past few decades, and I hope that as you read this book, you too will become excited about what such scientific skin care may offer you.

The world we live in is no longer the environmentally friendly place that it used to be. Industrial pollution has changed the atmosphere for the worse, and, besides having contaminants that directly affect our skin, other pollutants have deprived us of the protective qualities of the stratosphere, and so we are now constantly subjected to more intensive ultraviolet irradiation. In addition, we are experiencing exposure to the corrosive effects of excessive free radicals.

In this book we will first try to explain in simple terms what is happening to our environment as far as it affects our skin. Then we will explore what we should do by topical treatment methods to minimize, and even reverse, those adverse effects.

## ERNST

We have learned a great deal more about the chemistry of skin since Dr Fernandes wrote his first book, published in Japan, and I am happy to say that our new knowledge confirms and re-enforces the major precepts that he described in the first edition. However, much of this wealth of information lies untapped, whereas out-dated and even invalid ideas flood the market and the Internet to mislead people who want to know more about the science of skin care.

If you were to ask a number of specialists in skin care to give advice about the best way to treat skin, the chances are that you would get as many conflicting opinions. The main reason for this is that even 'specialists' in skin care tend to ignore the basic biochemical details of skin, and some therapists give advice about skin care without knowing any scientific principles. Some of the concepts they base their advice on will often be obscure and complex with only vague references to modern science. The concepts may even depend heavily on antiquated treatments, as though ancient people actually possessed the secrets for a healthy youthful skin! We know that those same ancient people got very wrinkled and pigmented if they went out into the sun, and they did not have any secret solutions to this age-old problem.

On the other hand, some people are easily swayed by current trends. They want to use only the most fashionable treatments, even though they know deep down that the manufacturing methods are more related to marketing, unsubstantiated medical claims, and media hype than reality. Others prefer to use products endorsed by famous personalities without being sure that the particular personality actually uses the product. In general there is another common belief that if a product is expensive it has to be more effective than more reasonably priced products. Modern skin care may certainly still use older techniques or time-honoured ingredients, but only when there are good scientific reasons to justify their use.

Equally, modern ingredients have to prove their real value before they should be recommended for general use. The fact that they are newly discovered does not mean that they are more effective than ingredients that have been used and proven for a longer time. In fact, we have to accept that there are certain essential chemicals that will always be necessary to protect our skin from light, free radicals, and the changes associated with ageing. These are chemicals found naturally in our skin. They have evolved with us as a species, and have functioned very effectively to protect us from sunlight over eons and will continue to do so into the future. Since these highly active molecules are continuously destroyed by exposure to sunlight, we need to diligently feed our skin with them. Modern science cannot change the basic chemistry or the evolved physiology of human skin. New discoveries merely change our ability to understand the intricate chemical pathways that are at work in the cells of the skin. By understanding what is happening, we can tailor our treatment more precisely and direct it more scientifically to make healthy young-looking skin.

Dr Fernandes and I have collaborated intensely to write this book together. He is in the fortunate position of having a laboratory that can make almost any product that he wishes, and he has spent many years building up a profound understanding of what happens to skin as it ages, and especially when we spend longer time exposed to the sun. We both want to understand precisely what the skin needs when it has been damaged. In the process of studying the skin in such intricate detail we have learned a lot of interesting, new, and surprising details that we would like to share with you, the reader of this book.

We will interpret the complex chemical details in as simple a language as we can. At the same time we also want to support the details with valid scientific research and evidence-based facts.

We hope this book will convey the knowledge that it is important to maintain a rational and scientific skin care habit. Thanks for joining us on this venture to try and rationalize skin care scientifically and help others to get a beautiful skin for a lifetime. When gaining all this knowledge, it is always important to remember that we are ultimately the beneficiaries or the victims of our habits. We can choose.

Des Fernandes
Ernst Eiselen
Cape Town & Perth
June 2014

# Chapter 02
# A PERSPECTIVE ON MODERN SKIN CARE BY DR DES FERNANDES

Skin care started several centuries ago when people realized that by rubbing fats on the skin they could make the skin feel smoother. Of course, the fats did not smell good and so clever people searched for ways of making emulsions that could be perfumed to mask the smell. From then on cosmetic creams remained virtually unchanged in principle through the centuries. As time went on, chemists tried to make creams similar to sebum, the natural oil of the skin. In the last quarter of the twentieth century a radical revolution in skin-care science started, but the science that introduced that change had started even earlier.

In the early twentieth century scientists discovered vitamin A and began the long journey to unravel its complex and essential role for maintaining normal healthy skin. They realized that healthy skin required adequate doses of vitamin A in the diet, but they thought that vitamin A could be supplied only through the diet.

Up until the 1930s almost everyone believed that ageing skin was a condition that occurred because of the passage of time and was inevitable. However, around this time, research workers were able to isolate vitamin A and found that it was an extremely sensitive family of molecules. Exposure to light destroys the vitamin's activity, as does exposure to air and water. The colour of vitamin A also changes from a healthy orange colour

*The chemical structure of vitamin A*

when it is active to a brownish colour when it has been degraded; research workers therefore wondered whether the natural vitamin A in our skin was safe when exposed to sunlight. By 1935 Drs Wise and Sulzberger[1] started to question whether old wrinkled skin was in fact vitamin A-deficient skin. They realized that the vitamin A in sunlight-exposed skin must be damaged by exposure to light, causing a chronic deficiency disease in that area, whereas sun-protected areas remained smoother for much longer. They pointed out that wrinkling did not occur as easily in skin that was protected from sunlight, but few people were prepared to listen to their message. It has many functions, among which are protection from the environment, prevention of drying out, and help in forming vitamin D.

The idea of treating this vitamin A deficiency in sun-exposed skin was ignored until quite recently. The very first report of trying to reverse aged skin by using vitamin A was in the middle 1950s. Elderly patients were given an ointment of vitamin A to rub on their leg skin for about six weeks.[2] At the end of the period there was some improvement. Six weeks is rather soon to notice changes in skin, and other doctors who may have tried to do the same were probably discouraged by the slow improvement. However, that was a landmark in human endeavours to control ageing. Never before (as far as we know) had any topical treatment been used to reverse ageing. The form of vitamin A used was retinyl palmitate, the predominant form of vitamin A found in the skin and stored in the liver. That was the beginning of the modern era of skin care and it was almost immediately forgotten!

In the mid 1980s people were amazed by reports that photoaged skin could be improved by using retinoic acid (vitamin A acid).[3] They then assumed that only retinoic acid was effective, but it has the usual disadvantage of irritating the skin and making it drier. Retinoic acid can make the skin very sensitive to the sun and alarmingly red.

Under normal conditions skin has very little retinoic acid in it, even though it is the form of the vitamin that acts on the DNA of the cell. That is quite a paradox, but scientists are beginning to

*"The idea of treating vitamin A deficiency in sun-exposed skin was ignored until quite recently."*

explain the detailed chemistry. We now know retinoic acid is an irritant molecule and the cell wall needs special receptors to take it into the interior of the cell rapidly. Sun-damaged skin is depleted of these receptors, so the retinoic acid remains outside the cell longer and irritates the skin. This is called a 'retinoid reaction'. The milder form of vitamin A, retinyl palmitate, was really the first form of vitamin A to be effective in rejuvenating skin[4], but was regarded as an inactive form of the vitamin.

Skin care took another leap in the wrong direction in the late 1980s when some doctors claimed that alpha hydroxy acids ('fruit acids', AHA) gave the same results as retinoic acid[5] but did not have the disadvantages and could be used even in the sun. Because the AHA products did not seem to irritate skin too much and did not have the irritant retinoid reaction seen with retinoic acid and other versions of vitamin A, they were thought to be superior, and so vitamin A declined in popularity. However, with the passage of time, the side effects of prolonged use of AHAs (without any support from vitamin A) led to dry irritated skin that lost condition. Moreover, AHAs do not have the rejuvenating effect of vitamin A.[6] The market started to recede from AHAs and vitamin A was revitalized – but this time as retinol, the alcohol form of vitamin A!

Then in the 1990s vitamin C became the next craze: 'Throw away the AHAs and vitamin A, because vitamin C is the one that really works!'[7] Ascorbic acid is an antioxidant and is normally found in skin. Not surprisingly, vitamin C is sensitive to sunlight, particularly UVA and blue light. Vitamin C is considered to be the dominant vitamin required to protect cell membranes from free-radical damage and is also necessary for the production of collagen. Another bonus is that vitamin C can minimise the sun damage to our pigment-making cells. Treatment of the skin with vitamin C was claimed to be superior to vitamin A, without the nasty side effects. People stopped using vitamin A products and clamoured for vitamin C, but they forgot that vitamin A and C each do different things in skin and are both sensitive to light, so we cannot drop the one in favour of the other without compromising the natural metabolism of the skin. Then beta-hydroxy acids (BHAs) took further lustre off the AHAs, and for a while became a fashion, mainly because they can be very useful for acne as well as photoageing.

At last, in the late 1990s people realized that in fact oxygen-derived free radicals were a real danger to our skin and they started another fashion. Now the rage is still for antioxidants, and the marketing teams of the major over-the-counter cosmetics have focused on them. With each year a newer and 'more powerful' antioxidant has come to the fore with claims that it is far more effective and makes everything else unimportant. Marketing departments ignore the fact that we need as many antioxidants as possible with various properties to deal with the varying conditions in the cells of our body. A single antioxidant will never be able to

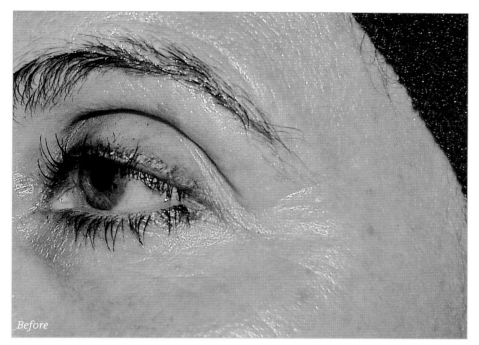

*Close up of "crow's feet" area before using skin care with vitamin A, C, and neuropeptides (matrixyl and leuphasil)*

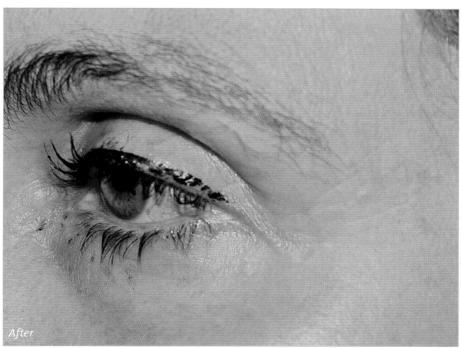

*After using vitamin A, C, E, B5 combined with neuropeptides for three months, the crow's feet lines are less evident and the skin is showing the benefits of vitamin A nutrition*

reverse ageing and can only have limited effects in slowing it down. Science tells us we need a broad spectrum of free-radical scavengers to protect our cells from lipid, nitrogen, carbonyl, and water-based free radicals.

The latest advance in skin care has been the introduction of peptides and other signalling molecules that carry messages to the cells in order to make collagen and elastin, among other things. We will see growth predominantly in this aspect of skin care in future. I have many projects under way using these molecules. I produced my first product with peptides in 1999. (the Environ Original eye gel and then the Environ Ionzyme C-Quence products)

This is a brief resumé just so that you can get some perspective on the work that I have been doing since 1987, when I first started to make skin-care creams with vitamin A, C, E, B5 and beta carotene in active doses.

The wheel has turned, and in recent years research workers have 're-discovered' retinyl palmitate and have shown what I pointed out very many years ago: the results from nourishing the skin with retinyl palmitate are indistinguishable from those achieved with retinoic acid.[8],[9] The main advantage is that they are easier to achieve without causing a retinoid reaction. As time goes by, we can expect more and more creams to introduce vitamin A as retinyl palmitate in effective doses. For your perspective on the matter, the doses of vitamin A used in the reports quoted above by Watson are lower than the initiating dose that I recommend. I believe that one needs much higher doses to get impressive changes and some anti-ageing effect. That is why it is essential to have a range of creams with a 'step-up' system starting at low doses and building up to the maximum permissible doses for a cosmetic product.

Let's now delve into the complex science of skin and discover the intricate essential role of vitamin A, which controls our destiny in amazing ways.

## REFERENCES

1. Wise F., S., MB Yearbook of Dermatol The Yearbook Publishers, Chicago, 1938. **282**
2. Reiss, F. and R.M. Campbell, *The effect of topical application of vitamin A with special reference to the senile skin*. Dermatologica, 1954. **108**(2): p. 121-8.
3. Kligman, L.H., *Photoageing. Manifestations, prevention, and treatment*. Dermatol Clin, 1986. **4**(3): p. 517-28.
4. David F. Counts, F.S., Julianne McBee, A. G. Wich *The effect of retinyl palmitate on skin composition and morphometry*. Journal of the Society of Cosmetic Chemists, 1988. **39**(4): p. 235-240.
5. Van Scott, E.J., C.M. Ditre, and R.J. Yu, *Alpha-hydroxy acids in the treatment of signs of photoageing*. Clin Dermatol, 1996. 14(2): p. 217-26.
6. Pi rard, G.E., et al., *Comparative effects of retinoic acid, glycolic acid and a lipophilic derivative of salicylic acid on photodamaged epidermis*. Dermatology, 1999. **199**(1): p. 50-3.
7. Darr, D., et al., *Topical vitamin C protects porcine skin from ultraviolet radiation-induced damage*. Br J Dermatol, 1992. **127**(3): p. 247-53.
8. Watson, R.E.B., et al., *A cosmetic 'anti-ageing' product improves photoaged skin: a double-blind, randomized controlled trial*. Br J Dermatol, 2009. **161**(2): p. 419-426.
9. Watson, R.E., et al., *Repair of photoaged dermal matrix by topical application of a cosmetic 'anti-ageing' product*. Br J Dermatol, 2008. **158**(3): p. 472-7.

# Chapter 03
# THE STRUCTURE OF SKIN

## BASIC ANATOMY

The skin is the biggest organ of the body and one of the most important. It has many functions, among which are protection from the environment, prevention of dehydration, and the formation of vitamin D. It is the organ housing senses of touch, temperature, and pressure.

Skin has a very specialized structure, and by understanding the composition and function of skin you can take better care of it. For this reason, the present chapter on the structure of the skin is worth reading carefully, even if it may seem complicated.

## STRUCTURE OF HUMAN SKIN

Skin consists of a number of different layers of cells and structural proteins embedded in a watery gel and cushioned by a soft layer of fat cells. Each layer has an important role. The combination of these layers lends strength and pliability to the skin. Human skin is in effect a highly effective and resilient laminated structure. In laminates each individual piece may not be very strong at all, but when properly glued together they form a resilient and, in the case of skin, a highly elastic and pliable overall structure.

The layers are called the epidermis, or outermost cellular layer, the dermis or deeper structural layer, and the subcutaneous fat layer. The skin is supplied with a vast network of small blood vessels that go as far as the upper

Composition of the skin

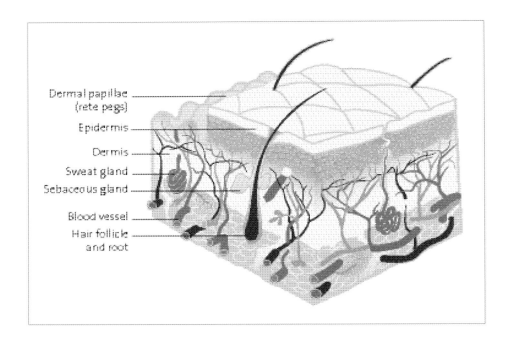

dermis, called the papillary dermis. The blood vessels and capillaries do not go into the epidermis.

The skin contains hairs with closely associated sebaceous glands, sweat glands and specialized nerve endings for all the different sensory functions.

As you can easily see by inspecting your own skin, there is tremendous variance in appearance and texture of the skin in different parts of the body. Look at your palms and the soles of the feet, where it is thickest and more rigid to endure the constant heavy wear and tear skin is subjected to here. In contrast, it is thinnest on the upper eyelids, where the skin needs to be most pliable. One advantage of this difference is that thick skin is less vulnerable to sun damage, but conversely, thin skin like that on the eyelids is very easily damaged even by indirect exposure to the sun.

Look at the skin carefully and you will see that the surface has fine hairs (except on the palms and soles of the feet) and is pitted by sweat glands (commonly called pores). In addition there are creases in the skin (called Langer's lines) that are determined by the elastic tension of the skin. These lines are very important to plastic surgeons, because scars placed in the direction of Langer's lines are far less noticeable than scars that cross them. A fine polygonal pattern (geometrical shapes of great variety) is visible on the very outer surface of the skin. This is the normal skin topography, but on damaged or scarred skin this topography or surface structure becomes altered, destroyed, or different to surrounding skin. The restoration of this pattern is one way of judging whether skin has regenerated normally after injury.

The skin changes over the years and we all hope to minimize changes associated with ageing. In babies it is smooth and dry and there are few blemishes. With the passage of time numerous changes occur: the skin gets hairy and pigmented marks develop, disturbing the evenness found in perfect juvenile skin. Healthy skin at four years of age is probably the best skin humans can ever have. The closer one can get skin to this state, the better. In adolescence hair grows under the influence of the sex hormones; moles (naevi) start growing (especially if skin is exposed to sunlight [1]), the skin gets oily, and acne occurs in many adolescents because of the secretion of testosterone and other factors. Pigment distribution changes and is much more susceptible to the influence of light. With the onset of maturity, wrinkles occur and various blemishes become aggravated, depending on the amount of exposure to sun and wind and so forth.

Then, as one ages, the skin starts to sag as it loses its elasticity and subcutaneous fat cushion. The skin also functions poorly and starts to give problems relating to the integrity of the laminated elements, to environmental protection, and to temperature control. These changes have always been considered unavoidable but this book will show how they may be avoided or minimized.

## THE MAJOR LAYERS OF THE SKIN

The skin consists of three major sections: a) the epidermis, which contains the most superficial (but mainly dead or dying) protective cell layers of the skin; b)

*"Changes due to ageing have always been considered unavoidable but this book will show how they may be avoided or minimized."*

*Skin close up*

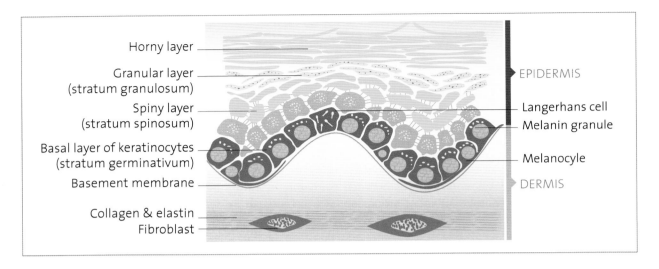

*Layers of the skin*

the underlying dermis that is an equally important living organ; c) a cushioning sub-cutaneous layer of highly organised fat cells, which provide the other two layers of the laminate with an elastic bed and support, while assisting in creating a surface volume so that the upper two layers are smoother and plumper.

## THE EPIDERMIS

The average thickness of the epidermis is only about 0.1 mm but it consists of five perfectly organized layers. The only actively living, dividing, layer is the basal layer. We'll look at the skin from the bottom where the cells are formed, and progress up to the surface where the cells are shed.

**I. THE BASAL LAYER** is the actively growing stem cell layer from which the epidermis develops. It is also the layer in which pigment is to be found. This layer is largely made up of stem cells, which are the origin of all the cells in the epidermis. These stem cells are considered undifferentiated (not yet specialized) cells, but they produce cells that can become differentiated (specialized) into keratinocytes. The individual stem cells in this layer divide into two (mitosis). A 'daughter' stem remains in the basal area, whereas her twin keratinocyte cell is pushed towards the surface of the skin. Keratinocytes are very important in the production of active molecules that control the formation of melanin, various collagens, and growth factors.

After dividing into two, the cells go into a resting phase and then start the cycle again. The process takes about 12 to 19 days. The cells are most vulnerable to UV light damage while they are dividing. From the formation of a cell in the germinal layer until it reaches the horny layer, where it is shed (desquamated), takes between 21 days to three months depending on the anatomical area of skin. The facial skin, for instance, grows very fast in contrast to skin on the back and on the legs, where the cells divide and grow much slower.

Because of the constant renewal of the skin, changes can occur in the chromosomes and abnormal cells may be formed. It's rather like Chinese whispers where the original message becomes more distorted the more times it is repeated. That is what happens to a small degree with our DNA. This process will be described in greater detail in a later chapter.

Fortunately the skin possesses special cells called Langerhans cells that are immunologically active and selectively destroy those cells with abnormal DNA, so reducing the likelihood of cancer formation. An intricate built-in gene-driven mechanism of programmed cell death called apoptosis also prevents abnormal cells from growing into skin cancers.

Beneath the basal layer is the basal membrane, which is permeable but divides the germinal layer from the dermis and allows it to be nourished by the dermis. If you look at the diagram of the skin you will notice that the junction between the epidermis and the deeper dermis is not smooth: it undulates and has projections which go deep into the dermis. Some of these are hair follicles, others are sweat glands but there are wider penetrations of the epidermis into the dermis (rete pegs or dermal papillae). These undulations stabilize the epidermis and fix it to the dermis by increasing the contact surface area between them. That also means that the epidermis is better nourished when there are well-developed dermal papillae, and becomes correspondingly less well nourished when the dermal papillae are flattened. Osmotic movement of molecules is more effective over a large surface area than a small one.

Between all the cells of the skin there is a minute space filled with fluid that is important for the nourishment of the cells of the skin. This fluid contains glycosaminoglycans (gel-like combinations of proteins and sugars), which are very important for our understanding of how the skin works. As the cell migrates to the surface it gets further and further from the nourishment, and so it dies. But there is a lot more to the story, as you will see.

II. THE SPINY LAYER (stratum spinosum) lies above the basal layer, and is so named because there seem to be many spiny projections called desmosomes from the cells connecting them to one another. With great magnification we can see that these 'spines' are really contact areas between cells. This means that the cells stick to each other and cannot easily be shed.

III. THE GRANULAR LAYER (stratum granulosum) lies higher up and is named for the granules that can be seen in these cells (The granules are kerato-hyalin (which is a pre-cursor of keratin) and membrane coating granules ) that will help to form the waterproofing barrier (see below). These are special protein-containing keratin structures and are not pigment granules.

> *"Because of the constant renewal of the skin, changes can occur in the chromosomes and abnormal cells may be formed."*

> "Skin colour depends on the number of blood vessels lying under the epidermis and the amount of pigment granules in the epidermis."

IV. **THE CLEAR LAYER** (stratum lucidum) is usually found only in the palms of the hands and the soles of the feet. The keratinocytes of the stratum lucidum do not feature distinct boundaries and are filled with eleidin, an intermediate form of keratin.[17]

V. **THE HORNY LAYER** (stratum corneum) is the layer where the cells lose their internal structure, become flattened, and are by this time filled only with keratin. They progressively lose their adhesions to each other so that they may be shed in an orderly fashion. In the stratum lucidum and the lowest horny layer there is a natural waterproofing barrier that helps prevent water loss and determines the permeability of the skin. This barrier consists of lipids in double layers that are extruded from the dying cells and accumulate in the intercellular space to build up the impermeability of the stratum corneum. This barrier is only 0.01 to 0.02 mm thick and is therefore very sensitive to external influences, and is particularly strained by frequent use of detergents, abrasive products, the sun, weather, and pollution. Damage to it can lead to increased loss of moisture and consequently to dry, scaly skin.

You can see that the horny layer (stratum corneum), our real protection from the outside world, is extremely thin, and we need to keep it as healthy as possible. Keeping it healthy implies two basic characteristics in this layer: the production of the best quality protein and very tight binding of the cells to each other and the layer below them. Destroying this layer by abrading it makes no sense at all. It does not improve skin quality; instead it temporarily weakens the skin's defences.

That is the basic structure of human skin, but of course we know that there are many variations in colour and texture and subtle individual variations in how skin reacts to the environment.

## SKIN COLOUR

Skin colour depends on the number of blood vessels lying under the epidermis and the amount of pigment granules in the epidermis. The pigment granules are melanin, which is manufactured by another set of cells (melanocytes) and then transported mainly to cells in the basal layer of the epidermis. The melanocyte is largely controlled by keratinocytes and melanocortin. Through their influence melanin is made by a complex chemical reaction in the melanocyte, which is heavily dependent on the essential enzyme tyrosinase. Tyrosinase acts on tyrosine to eventually produce melanin.

Surprisingly, there are only two different types of melanin: (a) dark brown-black melanin (eumelanin) and (b) yellowish-red melanin (phaeomelanin). These two melanins are responsible for the variety of colours of skin and hair. White hair has no melanin, whereas blond hair has traces of yellowish or brown melanin. The brown melanin (eumelanin) protects us from the sun—but its value should not be overstated, as the process of acquiring it causes damage to the skin. A tan is not a good way to build up protection from

*Variation of Melanosomes*

Melanosomes in dark skin are large and 88% are individual. Size 1.4μm. >200 melanosomes per cell. Degrade slowly and survive right into str.corneum.

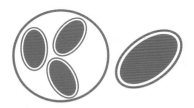

In Asian skin there is a mixture of about 62% large and single and 38% clusters size 1.34μm.

In pale skin 84% are smaller, clustered size 0.9μm. Usually clusters contain 2-10 melanosomes <20 melanosomes per cell. Degrade rapidly and have disappeared by stratum corneum and only melanin survives.

the sun, for even a good tan has an SPF value of only about two.

Eumelanin is, however, one of the most powerful free-radical scavengers and can deactivate free radicals caused by UV light. On the other hand, the reddish brown melanin (phaeomelanin) actually aggravates the effects of the sun, and that is why people with red hair tend to be more sun sensitive and have a greater risk of skin cancer.

Paradoxically, the number of melanocytes in both blond and black skin is about the same (about 1000-2000 per square millimetre). However, there is a variance in the production of the melanin pigment, and there are many more melanophores (cells that carry melanin) in dark-coloured skin. When you get a suntan, the melanocytes become more active and produce more pigment, but they, in themselves, do not increase

*Composition of the skin*

28 \ The structure of skin

in numbers.[2,3] However, the structural detail of melanin does differ between Caucasian, Asian, and African skin.

Each melanocyte is connected to surrounding keratinocytes by long fine tubules through which the melanin is transferred. The melanin usually lies on the surface of the nucleus and helps to protect DNA from harmful UV rays. In the melanocyte there is no visible pigment because the chemical precursors of melanin are colourless. Almost immediately after melanin has been synthesized it is transferred from the melanocyte to other cells and becomes brown or reddish brown. When UV irradiates the transferred melanin it becomes darker. In normal circumstances, an umbrella-like collection of melanin develops above the nucleus and protects the keratinocyte DNA.

Because of the melanin in our skin we all have some degree of natural sun protection. It is estimated that people with pale skin (Fitzpatrick type I) have a natural SPF of 1.00, type II skin is about 1.67, type III about 2.0 type IV-V may have an SPF of about 4, and type VI skin an SPF of about 10 to a maximum of about 14.[4]

## THE DERMIS

The dermis is the important living layer of the skin and is thicker (1-4 mm) but not as precisely organised as the epidermis. Hair follicles and sweat glands are derived from the epidermis but penetrate deep into the dermis. These epidermal elements serve as a source of epidermal cover when it is lost through injury or burns.

Collagen and elastin fibres are the main structures in the dermis with blood vessels, nerve endings, and fibroblasts lying in the interstitial fluid. As has been pointed out above, glycosaminoglycans (mainly hyaluronic acid) are an important constituent of the interstitial fluid. Glycosaminoglycans are mucoid polymer molecules capable of holding vast amounts of water (like a jelly), and that combination maintains the turgidity, or plumpness, of the skin.

The fibroblasts are the most important and the only permanent living cells of the dermis. They produce fibres and certain important chemical compounds. The only way to preserve a youthful skin is to activate or regenerate our fibroblasts.

Fibroblasts protect collagen by inhibiting collagenases,[5] and at the same time produce pro-collagenases, which destroy old collagen ready to be replaced.[6,7]

Fibroblasts are biochemically very active and respond to a number of hormones. On the surface membrane of the fibroblast there are a number of specific chemical receptors which selectively allow oestrogens, androgens, cortisone, and cholesterol to be attached to the membrane and from there to penetrate the fibroblast.

Vitamin A and especially vitamin C are essential for the production of healthy new collagen. Fibroblasts have tentacle extensions and are surrounded by interstitial fluid into which they secrete fibronectin and other proteins. Fibroblasts also contain organelles

*"Because of the melanin in our skin we all have some degree of natural sun protection."*

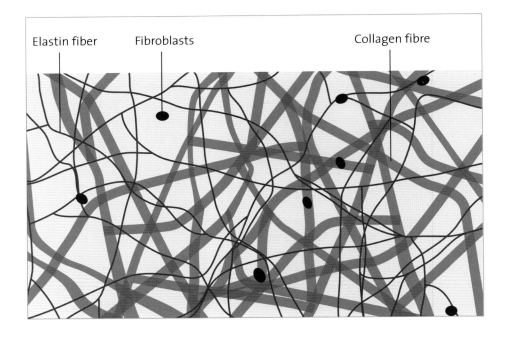

*The dermis consists of a few cells and many fibres embedded in a gel substance made of glycoseaminoglycans. The collagen fibres are in a lattice entwined with elastin fibres*

called hyalurosomes that are responsible for the production of hyaluronic acid (hyaluronan).[8] Hyaluronic acid is essential for retaining water in the interstitial matrix, and is therefore central to the maintenance of skin hydration, cell movement and the multiplying of cells in the dermis. Keratinocytes also produce hyaluronan (hyaluronic acid), which is important in maintaining epidermal moisture levels.[9, 10]

## HYALURONIC ACID

Hyaluronic acid is a highly visco-elastic polymer of glucuronic acid (a sugar) and glucosamine (an amine) and is therefore labelled a glycosamino-glycan (GAG). This acid is an important component of skin but has a high turnover rate. Hyaluronan has a major structural function to keep the dermis at optimum volume because of its amazing ability to hold water and cushion cells and structures in a protective gel. This property has an important shock-absorbing function that protects blood vessels and other structures. While we are not really aware of hyaluronan, we certainly notice when our skin becomes deficient in it: we get those bruises associated with thin skin and ageing. The skin becomes drier and fragile, breaking down easily. In the worst-case scenario, the bruises may actually kill the overlying skin in a condition called dermatoposis.

The production of hyaluronic acid, or hyaluronan, is initiated by cell-membrane receptors called CD44[11,12], which are stimulated mainly by vitamin A,[13] but also by medium-

**WHAT DOES A FIBROBLAST LOOK LIKE?**

The fibroblast is a long spindle-shaped cell with a round or oval nucleus. Fibroblasts have many functions. Including the production of:

- Many collagens
- Elastin
- A structural protein called fibronectin, important in the structure of collagen
- Laminin (which forms part of the basal membrane)
- Proteoglycans (which form the all-important sustaining gel of the inter-cellular matrix)

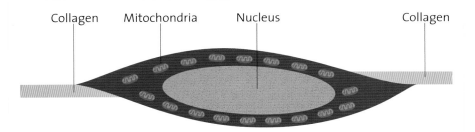

length fragments of hyaluronic acid.[14] Keratinocytes are also able to provide hyaluronan. On the other hand, UV damage causes the decay of CD44 receptors, and when our vitamin A levels in the skin are low, we have far fewer of them.[15] In turn, the CD44 receptor stimulates fibroblast hyalurosomes to produce hyaluronic acid.

We constantly need to replace hyaluronan, and the most effective way to do that is to ensure adequate levels of vitamin A in the skin. We cannot apply it topically because hyaluronic acid is a very large polymer and cannot penetrate skin even when relatively short fragments of hyaluronic acid are used. Research has shown that topically applied hyaluronan does not raise the CD44 levels or increase the hyaluronan in the skin. The only way to ensure delivery of hyaluronan to the deeper levels of the skin is either to use low frequency sonophoresis or to do microneedling of the stratum corneum to allow the hyaluronan to penetrate. Without the needling the hyaluronic acid molecules are too large to penetrate the skin. The combination of topical hyaluronan and retinoids after surface needling (using, for example, the Environ Cosmetic Roll-CIT) gives the best results.

## THE FORMATION OF COLLAGEN AND ELASTIN

First of all, pro-collagen, the precursor of collagen, is made and transported down the microtubules of the endoplasmic reticulum in the fibroblast and then excreted out of the

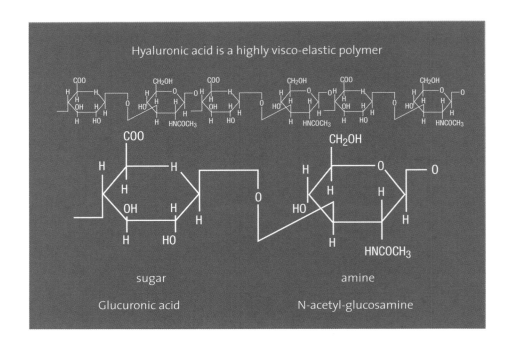

*This is a representation of the chemical arrangement of Glucuronic acid and n-acetul-glucosamine that is polymerised to create chains of hyaluronic acid that can combine with many times its weight of water.*

cell into the interstitial fluid, where it develops into collagen. The command system to produce collagen is in the chromosomes of the nucleus, but the production of collagen begins in the endoplasmic reticulum, where the messenger RNA from the nucleus creates effects on protein creation and continues into the Golgi apparatus, which is an organelle that completes the manufacture of collagen and other proteins which are then extruded. Fibronectin is a kind of glue that sticks the fibroblast to the collagen fibres, but in ageing, fibronectin diminishes and the fibroblasts become separated from the collagen. Laminin is another protein glue produced by the fibroblast that helps the dermis adhere to the epidermis. Collagen fibres become stiffer with age and are increasingly bound to other collagen fibres.

In order to prevent or delay the ageing process we must promote or maintain activity of the fibroblast. There are limited ways to do this. We can activate the fibroblast by: (1) supplying oestrogens; (2) supplying vitamin A derivatives; (3) altering the DNA; (4) using antioxidants; (5) supplying peptides or messenger molecules that instruct the fibroblast to make healthy collagen.

## HAIR

Almost the whole body is covered with hair of varying thickness. The upper eyelids, the penis, labia minora, the palms of the hands and the soles of the feet do not have hair. (The labia majora have coarse hair.)

Hairs are manufactured in indentations of the epidermis at the bases of which are little dome-shaped collections of cells from which the hairs are formed. Hair is a structure formed of compacted keratin, where the pigmentation is largely derived from melanin. Hair is basically dead tissue, and so cutting hair has no effect on its growth; shaving skin with fine hair will not make the hair become coarse. Hair has characteristic growth phases and rest phases. During rest phases hair loss increases, but these lost hairs eventually grow back. Typically, hair tends to go into a rest phase after pregnancy. Thinning hair at this time often causes distress in women, but the phase is normally temporary and the hair grows back. Hormonal changes besides pregnancy and emotional stress may increase the length of rest phases.

## SEBACEOUS GLANDS

Sebaceous glands are responsible for the oily coating of the skin and are usually attached to hair follicles. They discharge their secretions into the hair follicle through a narrow tract called the sebaceous follicle. The activity of the sebaceous glands is largely controlled by the male hormone dihydrotestosterone (DHT), which is a derivative of testosterone and is also implicated in male-pattern baldness. Testosterone is converted by alpha-reductase enzymes to DHT. Male hormone is naturally formed even in women by their ovaries and adrenal glands. Stress also plays an important part in stimulating the activity of sebaceous glands, as stress causes the secretion of more cortisone from the adrenal glands, which contributes to an increase of sebaceous gland activity. There are often large numbers of sebaceous glands on the forehead, nose, and cheek area, where blackheads tend to develop more readily. Areas where sebaceous glands are not associated with hair follicles are on the upper lip, inside the mouth, on the eyelids and breasts. [16]

## SWEAT GLANDS

Sweat glands are fine tubules derived from the epidermis that lie in a coil in the dermis and have a fine duct to the surface of the skin. The sweat is secreted onto the surface of the skin, evaporates, and consequently helps the body to lose heat. Sweating is essential to prevent the body from overheating. In addition to sweating for heat loss, perspiration can also be induced by psychological stimuli, but this sweating usually occurs on the palms and soles and sometimes the forehead and face.

One kind of sweat gland is called the apocrine gland, which secretes into the hair follicles in the armpit, the ear canal, the nipples, and ano-genital areas. The odour of axillary secretions becomes more intense as bacteria decompose it. The odour is not universally perceived as unpleasant and probably plays a large part in sexual attraction in some individuals.

## NAILS

Nails are a major characteristic of primates. They are closely related to hairs

*"Stress also plays an important part in stimulating the activity of sebaceous glands"*

but grow continuously and do not have resting phases. They grow at about 0.1mm per day (about a third of the growth rate of hair) and are affected by the seasons (growing slower in winter), age (growing fastest in the young adult), and disease. With stress, fevers, and some drugs the nails may become thinner, thicker, cracked, or grooved. They may be deformed and in some cases even be shed. In rare cases they are completely absent.

## SENSE ORGANS

There are numerous nerve endings that allow us to appreciate environmental stimuli such as temperature, pressure, and pain. Each one of these nerve endings has a particular structure related to its function. Very broadly one may say that there are no significant numbers of pain fibres in the epidermis, so needles that penetrate only the upper layers of the epidermis cannot be felt, as in 'cosmetic' needling.

## BLOOD VESSELS

The skin has probably the richest supply of blood in the body compared to all other organs. As a result, regulation of the blood flow is very important in controlling temperature and blood pressure. That is why some people blush, and at other times they may 'turn white' with rage or other emotions. It all depends on the amount of blood flowing through the skin at the time.

## LYMPH VESSELS

Lymphatic vessels are poorly defined but have an important role to play. They seem to be a passive type of drainage of the skin, as the flow of lymph is determined by gravity, massage of the skin, and other external physical forces. The lymphatic vessel walls are very flabby and easily collapse. These vessels start as blind capillaries in the interstitial spaces between the cells of the dermal papillae. They help to remove excess fluid, antigenic substances (that is, substances that stimulate an immune reaction), foreign molecules, and bacteria, and they help to preserve important proteins by getting them back into the venous circulation.

## SUMMARY

Skin consists of a number of layers that protect the body and give it its characteristic pliability. There are two main layers of the skin: the outermost epidermis is the 'waterproofing' which is concerned with physically protecting the body, and the innermost dermis provides the gel structure that cushions and nourishes the tissue, including the epidermis. The elastic and collagen framework for the skin is in this gel. The dermis is involved in temperature regulation through its extensive network of blood vessels.

In the epidermis the important cells are: (1) the keratinocytes that are the main growing cells of the skin and also serve as the pigment-carrying cells (melanophores); (2) the pigment-creating cells (melanocytes); and (3) Langerhans cells, the skin-patrolling cells regulating the immune response of the skin and thereby controlling the development of skin cancer and allergic states and the detection of microscopic invaders and foreign bodies.

The epidermis is critical for waterproofing the skin, and here the 'dead' layers play a crucial part. As the cells die they produce structures and chemicals that constitute a barrier called the horny layer, which is essential for maintaining the hydration of the skin. It also determines the softness and suppleness of the skin as well as keeping the skin free from infections. It is important for you to maintain this horny layer carefully so that your skin will not dry out.

The dermis has only one important cell: the fibroblast. The fibroblast is the key cell in maintaining or restoring a youthful appearance to the skin. However, structurally the dermis is almost totally made up of a hyaluronic-acid-based gel in which the elastin and collagen fibres are embedded. Hyaluronan is the pivotal functional and structural molecule of the dermis. We have to learn to maintain healthy levels of hyaluronan in the skin if we wish to age well.

## REFERENCE

1. Rampen, F.H. and P.E. de Wit, *Racial differences in mole proneness.* Acta Derm Venereol, 1989. **69**(3): p. 234-6.
2. Abdel-Malek, Z., et al., *Analysis of the UV-induced melanogenesis and growth arrest of human melanocytes.* Pigment Cell Res, 1994. **7**(5): p. 326-32.
3. Eller, M.S., K. Ostrom, and B.A. Gilchrest, *DNA damage enhances melanogenesis.* Proc Natl Acad Sci U S A, 1996. **93**(3): p. 1087-92.
4. Cripps, D.J., *Natural and artificial photoprotection.* J Invest Dermatol, 1981. **77**(1): p. 154-7.
5. Welgus, H.G. and G.P. Stricklin, *Human skin fibroblast collagenase inhibitor. Comparative studies in human connective tissues, serum, and amniotic fluid.* J Biol Chem, 1983. **258**(20): p. 12259-64.
6. Welgus, H.G., et al., *Human skin fibroblast collagenase: interaction with substrate and inhibitor.* Coll Relat Res, 1985. **5**(2): p. 167-79.
7. Ditre, C.M., et al., *Effects of alpha-hydroxy acids on photoaged skin: a pilot clinical, histologic, and ultrastructural study.* J Am Acad Dermatol, 1996. **34**(2 Pt 1): p. 187-95.
8. Sakai, S., et al., *N-methyl-L-serine stimulates hyaluronan production in human skin fibroblasts.* Skin Pharmacol Appl Skin Physiol, 1999. **12**(5): p. 276-83.
9. Sakai, S., et al., *Hyaluronan exists in the normal stratum corneum.* J Invest Dermatol, 2000. **114**(6): p. 1184-7.
10. Sayo, T., S. Sakai, and S. Inoue, *Synergistic effect of N-acetylglucosamine and retinoids on hyaluronan production in human keratinocytes.* Skin Pharmacol Physiol, 2004. **17**(2): p. 77-83.
11. Swweath, R.J. and D.C. Mangham, *The normal structure and function of CD44 and its role in neoplasia.* Mol Pathol, 1998. **51**(4): p. 191-200.
12. Lee, S.E., et al., *Stimulation of epidermal calcium gradient loss increases the expression of hyaluronan and CD44 in mouse skin.* Clin Exp Dermatol, 2009.
13. King, I.A., *Increased epidermal hyaluronic acid synthesis caused by four retinoids.* Br J Dermatol, 1984. **110**(5): p. 607-8.
14. Kaya, G., et al., *Hyaluronate fragments reverse skin atrophy by a CD44-dependent mechanism.* PLoS Med, 2006. **3**(12): p. e493.
15. Calikoglu, E., et al., *UVA and UVB decrease the expression of CD44 and hyaluronate in mouse epidermis, which is counteracted by topical retinoids.* Photochem Photobiol, 2006. **82**(5): p. 1342-1347.
16. K. R. Smith* and D. M. Thibotot1,*,†, Sebaceous gland lipids: friend or foe? Jake Gittlen Cancer Research Foundation* and Department of Dermatology,† Pennsylvania State University College of Medicine, Hershey, PA 17033

# Chapter 04
# THE SUN AND FREE RADICALS

The sun radiates numerous types of waves; most are absorbed to variable degrees by the ozone layer. Ultraviolet light is conveniently divided into A, B, and C rays. UVA and UVB rays are not completely absorbed and affect us directly by penetrating our skin. UVC, at present, stops at the stratosphere. The exception is in spring when over the south and north poles the UVC penetrates through the holes in the ozone layer.

There is increasing awareness around the world that all light, including computer screens, fluorescent lights, and energy-saving bulbs, radiates UV rays and may cause photoageing. Most important is the longer wavelength ultraviolet A (UVA) that can penetrate clouds, so it is just as important to protect your skin from UVA in countries with cool climates as in countries under a tropical sun. The role of blue light and infra-red light in ageing is also a possibility that needs exploration.[1] Sunbeds should never be used, as they give off mainly UVA rays and damage the skin and increase the chances for photoageing and skin cancer.[2]

The sun, in the correct amounts of exposure, is also very good for us! We know that if we do not get enough sunlight we feel depressed. Sunlight causes pigmentation of the skin, which forms a natural (though weak) barrier against photodamage and photoageing.

Sunlight is also essential to maintain adequate formation of vitamin D, which prevents softening of the bones (osteoporosis), depression, impaired immunity, and some cancers.[3] Vitamin D is made only by UVB rays, and there is growing interest in its importance in preventing cancer.[4] Humans should not however, expose themselves excessively to the sun, because paradoxically, UVA rays destroy vitamin D! It takes about 20 minutes to convert pre-vitamin D to active vitamin D, so the longest that one should stay in sunlight is about 20 minutes. However, never forget that it takes only a few minutes of sunlight in summer to generate enough vitamin D each day. In fact, pale skinned people (Fitzpatrick I, II and III) need only about half the time that it would take to make their skin pink (Minimal Erythema Dose or 'MED'),[5] in order to make vitamin D. Darker-skinned people (Fitzpatrick IV, V, and VI) have an important hindrance: melanin may prevent UVB rays from penetrating and it may take five times longer to start making vitamin D.

It is believed that vitamin D helps to reduce the chances of many cancers. The paradox is that we have to expose ourselves to the summer sun because the atmosphere significantly reduces the

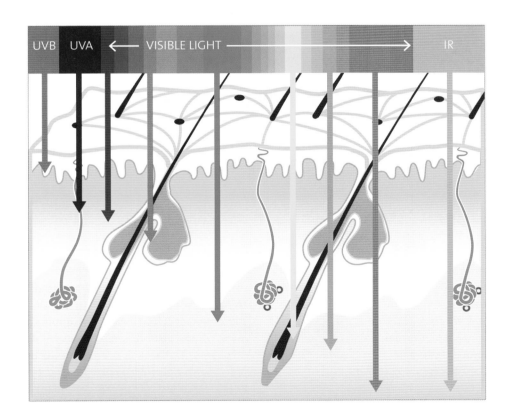

*UVB cannot penetrate the epidermis but UVA, visible and infra-red light progressively penetrate deeper into the dermis.*

*UVB loses its energy by the time it reaches the basement membrane. UVA can penetrate into the superficial dermis whereas IR (infrared) penetrates much deeper.*

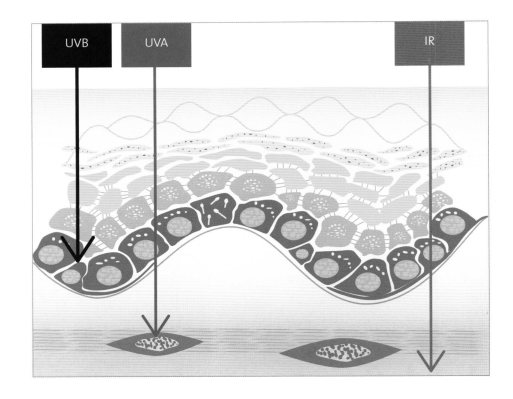

strength of UVB rays. From about mid-autumn through winter till mid-spring virtually no UVB rays can penetrate in those areas of the world north or south of about 25 degrees. That means that people living in the most populated areas of the world cannot make any vitamin D during that time—they can only get photoageing! That is also the time when influenza is common, and it is believed that this fact may reflect the impaired immunity associated with a low vitamin D status.[6]

Most sun damage is a result of exposure to sunlight on a daily basis – in shopping, driving, relaxing under an umbrella on the beach, and walking in the street. That is why it is important to wear protection all year round, not just when on holiday and soaking up the sun. By 'protection' we mean using sensible sun-wise clothes including wide-brimmed hats and using skin care that addresses the damage caused by light. Many people concentrate on blocking out the sun as though the sun itself is an enemy. In truth we need some sun exposure to have healthy levels of cancer-protective vitamin D. The shocking fact is that a significant part of our sun damage occurs when we are still teenagers. Damage sustained at this time accumulates and is not diminished by the passage of time.

Certainly there are negative effects of too much sun exposure, for instance both UVA and UVB deplete the skin of Langerhans cells, which are responsible for the immunity of the skin. The greatest danger of ultraviolet light, however, is the development of skin cancer caused by the formation of free radicals by UV light combined with the weakening of the local immune system in the skin, as mentioned above.

## THE PENETRATION OF UVA AND UVB RAYS

### UVA RAYS (A FOR AGEING)

UVA rays have a slow effect on the skin, and exposure to them manifests as increased pigmentation, that is, a tan. A tan is in effect a scar. It might, for a short time, be the most appealing scar we humans get, but it is a scar that never goes away.

UVA rays are longer wavelengths that are poorly absorbed by the ozone layer. They are approximately 500 to 800 times weaker than UVB but are about 1000 times greater in number.

UVA rays penetrate straight through the top layer of the skin and into the dermis. Their danger lies in the fact that there are no immediate visible signs of UVA damage; there is very little skin redness (erythema). Because these rays penetrate the bottom skin layer, or dermis, they can destroy cells essential to healthy skin. The damage is cumulative, which means that every day we add more damage; the older you are, the more damage you have accumulated and the more visible it becomes. The basic mechanism of UVA damage is mediated by free radicals. The energy of specific wavelengths of UVA rays is absorbed into a molecule such as vitamin A, and as a result this increased energy disrupts the molecule and renders it inactive. This effect causes a localized vitamin deficiency and is the main cause of photodamaged skin and the reason that vitamin A is so important in treating it.

### UVB RAYS (B FOR BURN)

UVB rays are shorter wavelengths and about 500 to 800 times stronger than UVA rays but about 1000 times less numerous. They penetrate only the top layer of the skin, the epidermis. They do not generally penetrate below the basal, growing, layer – depending on the thickness of the skin. Sunburn is the extreme, visible, physical damage of these rays, characterized by redness and blistering. UVB rays are essential for the manufacture of vitamin D in the skin, but at the same time UVB rays are the major cause of sunburn and skin cancer. UVB rays also help to destroy the vitamin A (retinyl palmitate) that is normally found in the stratum spinosum.

To reiterate: Do remember the visible effects of skin damage are cumulative, which means that the damage mounts up every day! Consequently, as we age the damage becomes increasingly more visible.

### UVC RAYS

UVC rays are the shortest and most powerful wavelength. At present, they

> "*UVA rays have a slow effect on the skin, and exposure to them manifests as increased pigmentation, that is, a tan.*"

do not enter our atmosphere, but they may already be reaching higher mountains, so UVC damage might be a risk to some mountain climbers. If these rays entered our atmosphere they would be extremely destructive to our skin. There are no cosmetics that can claim to have UVC ray protection because, at this time, there is no known specific, safe, protection from UVC rays.

## FREE RADICALS

Free radicals are essentially atoms that lack at least one electron, and in their quest for an electron to resolve the imbalance, they alter other atoms in other molecules, setting off a chain reaction that usually results in changes in thousands of molecules. This all happens in under a twenty-thousandth of a second!

In order to have an understanding of free radicals, we first have to consider some facts of chemistry. Everything in the universe is made up of matter, and in turn, matter is made up of atoms. In the centre of the atom is the nucleus, consisting of both positively charged protons and neutrons that do not have a charge at all. Circling the nucleus at a great distance are the electrons that carry a negative charge. For an atom to be stable it must have the same number of protons in the nucleus as electrons in orbit around the nucleus. The heavier the atom, the more orbits for electrons. In the outermost orbit, there should always be paired electrons for stability of the atom.

Light is made up of photons that are packets of energy, and the energy differs with the colour of light. An easy way to imagine this is as follows: a photon can strike an atom just at the moment that an electron is in the way, and that electron is shot off like a billiard ball into the surrounding environment. That means that this atom is now deficient of an electron, and has become a free radical searching for an electron to grab and restore the electrical balance.

Oxygen has eight protons and eight electrons and is given the chemical number eight. However, since two of the electrons in the outer circle are unpaired, single-atom oxygen is in a very unstable state, which is why oxygen is generally found in combinations of two atoms as $O_2$. The two atoms try to 'comfort each other' by sharing their electrons. However, these two electrons are very vulnerable to being dislodged, and this is why the oxygen molecule easily becomes a free radical.

A very clear example of this is the formation of rust when oxygen acts on iron. Only recently it has been realised that biological systems can also be 'rusted' by free radicals. Even more serious is the fact that we also know that iron itself can be a dangerous free radical in our bodies, and so the similarity to rusting becomes uncomfortably real. That's why we need healthy quantities of antioxidants all the time in our tissues.

Many people believe that free radicals are only bad for us, but that is not so! We also need free radicals as part of our natural energy production and as an important mechanism for defence against infections. Our white blood cells

> *"Only recently it has been realised that biological systems can also be 'rusted' by free radicals."*

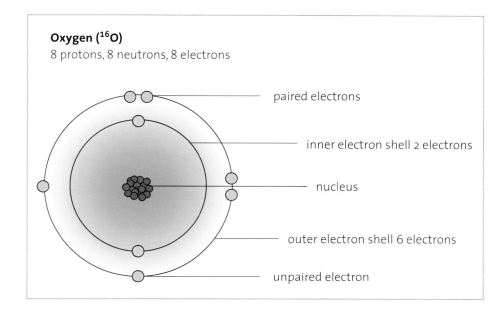

*The oxygen atom has two shells of electrons and the outer six electrons consist of two paired and two unpaired electrons. The unpaired electrons create the electrical instability that generates free radical activity.*

release free radicals to kill bacteria and viruses. If they release too many, we feel ill; if they release too few, we get infected. Without free radicals we would not have sufficient energy production in our mitochondria and we could not live.

## SOME INTERESTING FACTS ABOUT OXYGEN AND FREE RADICALS

Oxygen gas makes up one fifth of the air that we breathe at sea level; the higher one goes in our atmosphere the less oxygen there is. Parts of our atmosphere high above the earth have almost no oxygen. Only a very thin layer close to sea level has 20 per cent. We know that atmospheric oxygen consists of pairs of oxygen atoms ($O_2$). A single atom of oxygen has two unpaired electrons in the outer circle and is a free radical called a 'singlet oxygen radical'. Two percent of the oxygen taken up by the body is converted to toxic free radicals, so think again before ever having an oxygen-based facial.

Free radicals cannot be avoided, but their numbers are boosted by exposure to smoking, pollution, and UV rays. They say that our greatest exposure to free radicals is through smoking; one puff of cigarette smoke is reputed to hold over three trillion free radicals. People who do a lot of strenuous exercise have more free radicals in their body than someone who is inactive. We should beware of excessive exercises without adequate replacement of the antioxidant brigade.

While most free radicals are oxygen based, there are others such as nitrogen, chlorine, and carbonyl radicals, which are derived from proteins. By snatching

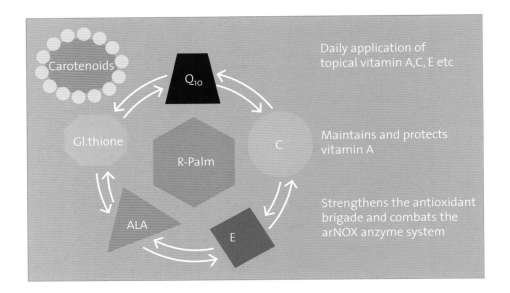

*This is the antioxidant network of the skin which comprises Glutathione, Coenzyme Q10, vitamin C, vitamin E and Alpha Lipoic Acid which recycle each component's antioxidant activity. This is fortified by carotenoids and altogether, the antioxidant network and carotenoids protect vitamin A in the skin.*

an electron from its neighbour, a free radical can set off a chain reaction that eats away at cell membranes and damages the DNA of the cells.

Our bodies produce an enormous number of free radicals during normal biological processes, and we have ways of handling them. The trouble starts with uncontrolled oxidation caused by free radicals. For example, when we have an infection and the body over-responds, an excessive number of free radicals are produced by our white blood cells to try to destroy the bacteria or virus. This response causes an overspill of free radicals which then makes the person feel ill. Free radicals distort a cell's capacity to reproduce itself accurately. They damage DNA, weaken cell membranes, alter collagen, fracture elastin fibres, change biochemical compounds within the body, and can even kill the cells outright.

There are both water-based free radicals and fat-based free radicals. Free radicals particularly affect the lipid (fatty) substances in our cells. These lipid substances make up the greater portion of the membranes that surround our cells in much the same way that the skin surrounds our body and protects the internal organs. These membranes also control the movement of nutrients and waste products in and out of the cells. When these fatty substances are attacked by free radicals, the membranes can no longer work properly. If the damage is extensive, the cell either dies or becomes distorted and vulnerable to germs and viruses that cause diseases. In their severest form free radicals can contribute to skin cancer. Damage by free radicals has been linked to premature ageing and wrinkling. They damage collagen and elastin, thereby causing the skin to become less resilient and more wrinkled. Damage to the DNA of keratinocytes is responsible for mutations that cause solar keratoses and cancer. Pigmentation occurs when melanin production is altered at the DNA level. Melanin is an extremely powerful free-radical quencher but paradoxically is not safe to apply to the skin surface.

Fortunately, excess free radicals can be inhibited to a certain extent by antioxidants. The best-known antioxidant compounds are vitamin E, vitamin C, and beta-carotene, all naturally found in skin. Antioxidants are provided by vegetables and fruit in our diet, which neutralize the radicals in our bodies by supplying the necessary electrons. When we go out into the sun we get an extra dosage of free radicals from UV rays, and consequently our antioxidants are overwhelmed, and we need more of them applied every day to the sun-exposed areas, ideally starting at an early age.

Vitamin C (ascorbic acid), besides being an antioxidant in the water phase of the body cells, is essential for protection of the lipoproteins of the cell membranes because it is important in 'reconditioning' vitamin E radicals created when vitamin E quenches free radicals. Vitamin C allows vitamin E to become an antioxidant once again.[7,8] These special antioxidant vitamins potentiate sunscreens, and their advantage is that they are absorbed into the skin cells and cannot be washed off when swimming or sweating.[9-11]

Research indicates that the addition of antioxidants is far more effective than sunscreens in protecting skin from cancer.[12] A very important point to bear in mind is that oral antioxidants in adequate doses are powerful ways to reduce free-radical-mediated UV damage in skin.[13] Interestingly, children may prepare for sun exposure by eating tomato-rich food such as tomato sauce or tomato soup![14, 15]

Note that not all skin damage caused by UV light is mediated through the creation of free radicals. The absorption of the energy of the UV photon is quite different and will be covered later.

Alpha-lipoic acid, glutathione, and coenzyme Q10 together with vitamin C and E are called the antioxidant network because they are capable of reactivating each other.[16] We should eat vegetables and fruits that are highly coloured because they contain rich antioxidant nutrients for us.[17-22] Research reveals that the Mediterranean diet with its characteristic distribution of different food groups and specific food types, may protect from melanoma.[23] Our diet may also be an excellent way to reduce age-related degeneration of the brain,[24-26] especially when we include nuts, fruits, and spices.

It is important to understand the mechanisms of damage from UV rays and realize that our lifestyles generally expose us to harmful amounts of them. The best way to counter this damage safely is to wear proper clothes and eat nuts, fruits, and vegetables.[27] These measures will automatically make the skin more photoresistant and minimize photoageing, and at the same time they will have enormous value in reducing

*"Research indicates that the addition of antioxidants is far more effective than sunscreens in protecting skin from cancer."*

age-related degeneration of our body as well as the brain.

## REFERENCES

1. Godley, B.F., et al., *Blue light induces mitochondrial DNA damage and free radical production in epithelial cells.* J Biol Chem, 2005. **280**(22): p. 21061-6.
2. Sinclair, C., *Vitamin D – an emerging issue in skin cancer control. Implications for Public Health practice based on the Australian experience.* Recent Results Cancer Res, 2007. **174**: p. 197-204.
3. Holick, M.F., *Vitamin D: a D-Lightful health perspective.* Nutr Rev, 2008. **66**(10 Suppl 2): p. S182-94.
4. Holick, M.F., *Vitamin D and sunlight: strategies for cancer prevention and other health benefits.* Clin J Am Soc Nephrol, 2008. **3**(5): p. 1548-54.
5. Cripps, D.J., *Natural and artificial photoprotection.* J Invest Dermatol, 1981. **77**(1): p. 154-7.
6. Cannell, J.J., et al., *Epidemic influenza and vitamin D.* Epidemiol Infect, 2006. **134**(6): p. 1129-40.
7. Thiele, J.J., et al., *The antioxidant network of the stratum corneum.* Curr Probl Dermatol, 2001. **29**: p. 26-42.
8. Guo, Q. and L. Packer, *Ascorbate-dependent recycling of the vitamin E homologue Trolox by dihydrolipoate and glutathione in murine skin homogenates.* Free Radic Biol Med, 2000. **29**(3-4): p. 368-74.
9. Hanson, K.M. and R.M. Clegg, *Bioconvertible vitamin antioxidants improve sunscreen photoprotection against UV-induced reactive oxygen species.* J Cosmet Sci, 2003. **54**(6): p. 589-98.
10. Oresajo, C., et al., *Complementary effects of antioxidants and sunscreens in reducing UV-induced skin damage as demonstrated by skin biomarker expression.* J Cosmet Laser Ther, 2010. **12**(3): p. 157-62.
11. Pinnell, S.R., *Cutaneous photodamage, oxidative stress, and topical antioxidant protection.* J Am Acad Dermatol, 2003. **48**(1): p. 1-19; quiz 20-2.
12. Beani, J.C., *[Enhancement of endogenous antioxidant defenses: a promising strategy for prevention of skin cancers].* Bull Acad Natl Med, 2001. **185**(8): p. 1507-25; discussion 1526-7.
13. Cesarini, J.P., et al., *Immediate effects of UV radiation on the skin: modification by an antioxidant complex containing carotenoids.* Photodermatol Photoimmunol Photomed, 2003. **19**(4): p. 182-9.
14. Stahl, W. and H. Sies, *Carotenoids and protection against solar UV radiation.* Skin Pharmacol Appl Skin Physiol, 2002. **15**(5): p. 291-6.
15. Afaq, F., V.M. Adhami, and H. Mukhtar, *Photochemoprevention of ultraviolet B signaling and photocarcinogenesis.* Mutat Res, 2005. **571**(1-2): p. 153-73.
16. Packer, L. and G. Valacchi, *Antioxidants and the response of skin to oxidative stress: vitamin E as a key indicator.* Skin Pharmacol Appl Skin Physiol, 2002. **15**(5): p. 282-90.
17. Harman, D., *The ageing process.* Proc Natl Acad Sci U S A, 1981. **78**(11): p. 7124-8.
18. Roy, S., et al., *Anti-angiogenic property of edible berries.* Free Rad Res, 2002. **36**(9): p. 1023-31.
19. Liu, M., et al., *Antioxidant and antiproliferative activities of raspberries.* J Agric Food Chem, 2002. **50**(10): p. 2926-30.
20. Kandaswami, C., et al., *The antitumor activities of flavonoids.* In Vivo, 2005. **19**(5): p. 895-909.
21. Ames, B.N., L.S. Gold, and W.C. Willett, *The causes and prevention of cancer.* Proc Natl Acad Sci U S A, 1995. **92**(12): p. 5258-65.
22. Khachik, F., et al., *Chemistry, distribution, and metabolism of tomato carotenoids and their impact on human health.* Exp Biol Med (Maywood), 2002. **227**(10): p. 845-51.
23. Fortes, C., et al., *A protective effect of the Mediterranean diet for cutaneous melanoma.* Int J Epidemiol, 2008. **37**(5): p. 1018-29.
24. Joseph, J.A., B. Shukitt-Hale, and L.M. Willis, *Grape juice, berries, and walnuts affect brain ageing and behavior.* J Nutr, 2009. **139**(9): p. 1813S-7S.
25. Willis, L.M., B. Shukitt-Hale, and J.A. Joseph, *Modulation of cognition and behavior in aged animals: role for antioxidant- and essential fatty acid-rich plant foods.* Am J Clin Nutr, 2009. **89**(5): p. 1602S-1606S.
26. Jurenka, J.S., *Therapeutic applications of pomegranate (Punica granatum L.): a review.* Altern Med Rev, 2008. **13**(2): p. 128-44.
27. Joseph, J., et al., *Nutrition, brain ageing, and neurodegeneration.* J Neurosci, 2009. **29**(41): p. 12795-801.

# Chapter 05
# PHOTOAGEING

I believe that photoageing should be classified as a chronic skin disease caused by light. My reason is that if people recognize it as a disease then they will realize that we need a scientific 'medicine' to treat this disease that affects virtually the whole world. Most outward signs of ageing skin result from cumulative everyday sun exposure. It begins in infancy and continues throughout life. It can take from ten to 40 years for the damage to manifest itself.

Photoageing has increased in the past few decades because of a change in our lifestyles. Prior to the 1920s it was uncommon for middle and upper class people to tan willingly because only labourers had tans. Then Coco Chanel started the fashion among the rich in the West to wear stark white clothing accentuated by a rich brown tan. A tan then meant that the person had the financial means to relax in the sun, while the workers stayed inside the factories and were consequently quite pale! Today many people are even more affluent and leisured – which usually means that they can spend even more time outdoors in the sun. As people are also living longer and now prefer to live in warmer, sunnier areas, sun damage is almost inevitable.

Photoageing tends to start in the teens, is clinically suspected in the 20s, and is clearly established in the 30s in Western people. Western people walk around with a constant natural sun protection factor (SPF) of about two, whereas Japanese people, for example, have an average of three to four SPF. [1] For this reason the first signs of photoageing in Japanese people usually only appear after the 30's.

## THE SIGNS OF PHOTOAGEING ON THE SKIN ARE:

❶ **IRREGULAR PIGMENTATION** comes from stimulated melanin production as a result of free-radical action and the release by keratinocytes of some active chemicals. Freckles are more common in women, as female hormones may aggravate them. Less commonly we see depigmented areas, probably the result of damage to the melanocytes. Pigmentation problems can occur even in dark-skinned people and is one of their biggest cosmetic complaints.

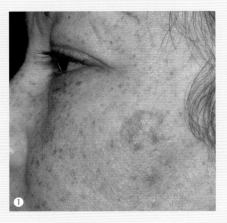

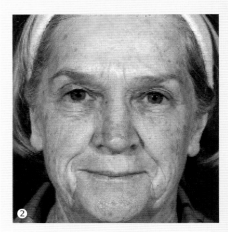

❷ **WRINKLING** is caused by the loss of collagen and elastin in the dermis and is aggravated by dehydration of the skin. It is an important cosmetic problem especially in the West, and much money is spent to counter or correct wrinkling.

❸ **LAXITY** occurs when the skin does not have sufficient elastic recoil to counteract the effects of gravity; so the skin sags, especially as one gets older and loses fat under the skin.

❹ **ELASTOSIS**, known also as 'chicken skin' or 'goose flesh' or 'turkey neck' on the neck and chest area, is due to clumps of elastin materials. It seems that the elastin fibre is fractured by UV light, and when it snaps it rolls into little balls. This sun damage is permanent and little can be done to heal it. Elastosis is often associated with red discoloured skin from telangiectasia of the blood vessels.

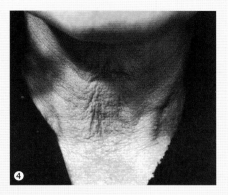

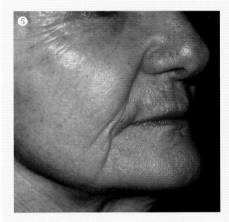

❺ **SALLOW SKIN** appears when blood vessels are damaged and less numerous; skin takes on a pale yellow appearance caused by the colour of the collagen below the epidermis.

❻ **ROUGHNESS** is evident when the skin becomes dry because of damage to the waterproofing barrier and the thickening of the horny layer of the epidermis. It seems that the purpose of the thicker horny layer is to create a thicker waterproofing barrier. This feature is more common in Caucasian skin and is only seen as a result of very severe photodamage in Japanese people, for instance. Different types of skin may have different reactions.

❼ **SOLAR (ACTINIC) KERATOSES** or 'sun spots' (or keratoses), may eventually occur due to damage of the DNA of irradiated cells, which then grow as a clone of abnormal cells that may ultimately show up as keratoses or skin cancers. Normally these cells would be removed by an effective immune system soon after they develop. However the Langerhans cells are deactivated by solar irradiation, the defective cells can thrive and create a clone or group of identical cells that slowly multiply and eventually become visible.

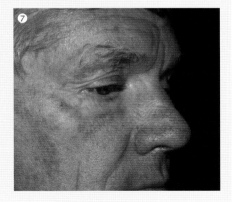

**❽ SOLAR COMEDONES** are large 'blackheads' (comedones) in the oily areas of the face, that is, on the forehead and nose, resulting from chronic severe sun damage. They are caused by an excessive amount of loose corneocytes that mix with sebum and clog up the follicles. Outdoors people, like professional fishermen with continuous severe sun exposure, will develop them.

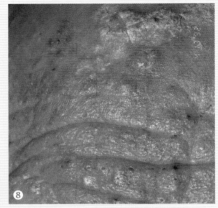

**❾ BASAL CELL PAPPILOMAS** (seborrhoeic keratoses) are usually brown to pink marks on the skin with a crusty top. It is not certain if they are signs of chronic photodamage, or areas where the DNA is no longer functioning correctly due to the shortening of telomeres and the arrest of normal cellular growth.[2] Since these lesions are also found on areas that have never been exposed to sunlight, photodamage may or may not play a part in their formation. It may, however speed up their growth. They form part of the signs associated skin ageing.

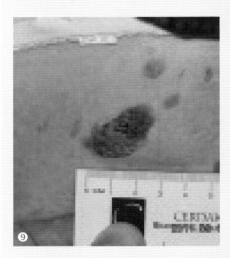

**❿ SKIN CANCER** develops in people with chronic sun damage. This can either be a basal cell carcinoma or a squamous cell carcinoma. Melanoma does not have a clear cut tie to photo-damage because it is often found in totally protected areas and some people attribute melanoma to low levels of vitamin D. [3, 4]

These are the outward signs of the damage that UV light in particular makes to skin, but we need to know how this happens by studying the chemical changes.

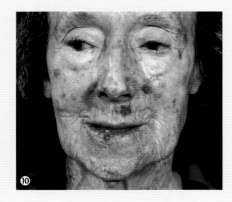

# THE CHEMISTRY OF PHOTOAGEING

We all start off with young skin and then, as time goes by, we get wrinkles and pigmentation, not because of the passage of time but because of the time spent in sunlight. The basic change in skin is caused by light entering the skin and then interacting with certain chemicals. This interaction can be seen either on the molecular or the atomic level.

On the molecular level light energy is absorbed into a molecule (a chromophore), and then such molecules respond in several ways. Some transfer the energy into light and so the chromophore emits light and becomes fluorescent. Others absorb the energy and radiate heat, or the energy is used to alter the structure of the molecule, as happens with vitamin A. The most important chromophore in skin is vitamin A as retinyl palmitate (RP), a molecule extremely sensitive to light. When vitamin A absorbs the energy of photons in the range of about 325 to 334nm, vitamin A activity is lost.[5] Interestingly enough, most sunscreens do not adequately protect retinyl palmitate molecules in the skin. There are many other important molecules that are damaged in the same way, for example, DNA, vitamin C, and vitamin B12.

On the atomic level, UVA rays instigate damage by electron changes that result in free-radical chain reactions. UVA rays predominantly cause photoageing because UVB rays cannot penetrate the epidermis and get into the dermis. UVA, visible light, and infrared rays can enter the dermis. As a general principle, the rays on the ultraviolet side tend to be damaging whereas the rays on the infrared side of the visible spectrum tend to be healing and beneficial to skin. UVA rays in general cause excitation of electrons and the creation of free radicals. However, there are important chromophores for UVA.

UVA can also activate genes at lower doses than that required to induce erythema. The best known chromophores for UVA are melanin, vitamin A (predominantly retinyl palmitate), [6] NADH (nicotinamide adenine dinucleotide), glutathione, vitamin D, riboflavin and tryptophan.[7, 8] Melanin continues to absorb all light forms without damage to the cell, but vitamin A is altered in its ring structure and becomes inactive and cannot be reconstituted; it also gives rise to free radicals.[9] UV light may also assist isomerization of all trans-retinoic acid to cis-retinoic acid[10] (that is, the changing of one molecule into another where the atoms are just shifted into different positions, retaining the same formula, but behaving differently as a chemical).

NADH is important as a source of energy and is altered by absorbing energy, when it can no longer function normally.[11] Glutathione is depleted by UVA and near visible blue light (405 nm–violet), as well as UVB 302 and 313 nm.[12]

Paradoxically, vitamin D is also sensitive to light and photodegrades easily once it has been formed in the skin. [13] For that reason people should not stay in the sun much longer than 20 minutes if they are intent on making vitamin

"On the atomic level, UVA rays instigate damage by electron changes that result in free-radical chain reactions."

D. This degradation is possibly also a protective feature because it prevents toxic levels of vitamin D.[13] Of course, we also have to recognize that the precursor of vitamin D is protective because it absorbs UVB rays. Vitamin D is also a sunscreen because it absorbs UVA rays.

Riboflavin and tryptophan can both absorb UVA and increase the formation of free radicals.[14] Free radicals themselves are responsible for lipid peroxidation,[15] the process in which free radicals 'steal' electrons from the lipids in cell membranes, resulting in cell damage.

UVB rays may penetrate only to the basal keratinocytes, but they can also induce changes in the release of chemicals from the keratinocytes that have profound effects, not only on the melanocytes, but also in the dermis. An example would be the release of matrix-metallo-proteinases (MMPs); powerful enzymes that can digest collagen (collagenases) and elastin (elastases). [16] UVB affects basal keratinocytes that then release precursors of MMP, and activators convert MMPs into the active MMP. A complex system of inhibitors of MMP release exists so that the skin does not automatically digest itself. One of the significant inhibitors of MMP release is vitamin A, but unfortunately when skin is exposed to sunlight, the vitamin A (largely as retinyl palmitate) is destroyed, and consequently the conversion of pre-MMP to active MMP is no longer fully inhibited.[17] The loss of retinyl palmitate in skin allows MMPs to be released to destroy collagen and elastin. Many complex chemical changes occur, and I believe it is extremely important to understand the chemistry of photoageing in order to then understand the mechanisms to treat it.

## THE CHROMOPHORES FOR UVB

The chromophores that absorb UVB are melanin, DNA, urocanic acid, vitamin E, 7-dehydrocholesterol (pre-vitamin D), and advanced glycation end products (AGEs). Melanin does not in general pose a problem for us as it absorbs the energy and will only create heat, and in fact absorbs any free radicals as well as chelating heavy metals.

DNA absorbs UVB at about 260 nm. This damage deserves special attention[18-20] because, although only the cells in the epidermis are affected, it is unfortunately a precipitating cause for skin cancer. Urocanic acid is among the normal oils secreted by the skin and acts as a natural sunscreen, but when it absorbs UVB energy, it becomes cis-urocanic acid, which promotes suppression of the immune system. [21-23] Vitamin E becomes deactivated by absorbing UVB rays.[24, 25] Because 7-dehydrocholesterol (the forerunner for vitamin D) absorbs UVB energy, it is converted into vitamin D, a process that takes about 20 minutes in the keratinocytes. Advanced glycation end products, or AGEs, accumulate on long-lived skin proteins such as elastin and collagen as a consequence of glycation. AGE proteins collect in the nucleus as well as the cytoplasm. Proteins modified by AGE can damage DNA with strand breaks from increased free radical activity as well as direct electron transfer between photoexcited AGE and DNA. So as we age we produce potent photosensitizers that make us age even more and cause more DNA damage![25] In addition, many enzymes are damaged or their action is promoted,[26, 27] and either way cellular changes occur. Mast cells in the dermis are increased in chronically sun-exposed skin,[28] resulting in higher levels of allergy sensitization and inflammation.

I believe the most important chemical component of photoageing is the depletion of retinyl palmitate (RP), which is the storage form of vitamin A in the skin and an effective cosmetic agent.[29-33] Many clinicians are unaware of the chemical changes that are involved in photoageing, and think that sunscreens are the best way to prevent it. They also believe that glycolic acid and other alpha-hydroxy acids (AHA) are the best way to treat it. Alpha-hydroxyl acids have no physiological role

> "Essentially, light enters the skin and certain wavelengths can damage it on the molecular or on the subatomic level because of the photon energy."

## THE SIGNS OF PHOTOAGEING IN DARK SKIN ARE:

Now that we have an idea of the chemistry of photodamage it should be interesting to see the tie-up to the clinical signs of photoageing.

❶ **IRREGULAR PIGMENTATION** arises primarily from free-radical damage to DNA from both UVA and visible blue light as well as from direct damage from UVB.[37] This damage causes the release of signalling molecules that induce the melanocyte to produce melanin. Vitamin A depletion at the same time means that pigmentary disorders can develop. We do not know exactly why retinoids are effective in controlling pigmentation disorders,[38, 39] but we do know that retinyl palmitate accumulates in the mid-epidermis [40] and acts as an umbrella to protect the cells below by absorbing UV rays that would otherwise cause damage and pigmentation. While dermal vitamin A may be relatively well protected by darker skin types, it could be that the loss of epidermal vitamin A, while it is acting as a sunshield, is instrumental in allowing pigmented marks to develop. No doubt the situation is aggravated by the loss of vitamin C and other antioxidants due to photodamage;[12] female hormones also somehow aggravate this situation.

*Before*

*After*

in photodamage and very limited roles in cellular physiology with the notable exception of lactic acid.[34]

While it is certain that sunscreens go a long way in preventing sun-damage, they do not actually treat the fundamental cause of photoageing and may aggravate the free radical challenge to skin.[35] Light consists of a whole spectrum of photons, which are 'packets of energy' that vary according to their wavelength. Essentially, light enters the skin and certain wavelengths can damage it on the molecular or on the subatomic level because of the photon energy. ***Most people believe that the damage is done only by UV light, but in fact even green, blue, and violet light can damage cells.[36] We should therefore also look at visible light as a potential cause of melasma.

## SUMMARY

This chapter has given you a brief idea of the photoageing eventually caused by the chemical changes that occur when we go out into sunlight. You can see that vitamin A depletion is central in the development of all the signs of photodamage. It seems that everywhere one looks Vitamin A is playing an important role in maintaining healthy skin. Thus, when we are deprived of it, the cells suffer and we get the signs of photoageing. To prevent or treat photoageing it is logical to design a therapeutic regime that restores damaged chromophores like vitamin A and C. We will go into this in more detail in the next few chapters.

## REFERENCE

1. Cripps, D.J., *Natural and artificial photoprotection*. J Invest Dermatol, 1981. **77**(1): p. 154-7.
2. Nakamura, S. and K. Nishioka, *Enhanced expression of p16 in seborrhoeic keratosis; a lesion of accumulated senescent epidermal cells in G1 arrest*. Br J Dermatol, 2003. **149**(3): p. 560-5.
3. Garland, F.C., et al., *Occupational sunlight exposure and melanoma in the U.S. Navy*. Arch Environ Health, 1990. **45**(5): p. 261-7.
4. Gambichler, T., et al., *Serum 25-hydroxyvitamin D serum levels in a large German cohort of patients with melanoma*. The British journal of dermatology, 2013. **168**(3): p. 625-8.
5. Berne, B., Nilsson, M., Vahlquist, A. , *UV Irradiation and Cutaneous Vitamin A: An Experimental Study in Rabbit and Human Skin.*, J. Invest Dermatol, 1984 **83**: p. 401-404.
6. Fernandes, D., *Understanding and treating photoaging*. Manders, E.K., and Peled, I.J. (Eds.), 2004. **Aesthetic surgery of the face**(Taylor and Francis, Abingdon, United Kingdom): p. 227-240.
7. Tedesco, A.C., L. Martinez, and S. Gonzalez, *Photochemistry and photobiology of actinic erythema: defensive and reparative cutaneous mechanisms*. Braz J Med Biol Res, 1997. **30**(5): p. 561-75.
8. Sorg, O., et al., *Spectral properties of topical retinoids prevent DNA damage and apoptosis after acute UV-B exposure in hairless mice*. Photochem Photobiol, 2005. **81**(4): p. 830-6.
9. Fu, P.P., et al., *Photodecomposition of vitamin A and photobiological implications for the skin*. Photochem Photobiol, 2007. **83**(2): p. 409-24.
10. Kunchala, S.R., T. Suzuki, and A. Murayama, *Photoisomerization of retinoic acid and its influence on regulation of human keratinocyte growth and differentiation*. Indian J Biochem Biophys, 2000. **37**(2): p. 71-6.
11. Hipkiss, A.R., *Mitochondrial dysfunction, proteotoxicity, and aging: causes or effects, and the possible impact of NAD+-controlled protein glycation*. Adv Clin Chem, 2010. **50**: p. 123-50.
12. Fuchs, J., et al., *Acute effects of near ultraviolet and visible light on the cutaneous antioxidant defense system*. Photochem Photobiol, 1989. **50**(6): p. 739-44.

13. Webb, A.R., B.R. DeCosta, and M.F. Holick, *Sunlight regulates the cutaneous production of vitamin D3 by causing its photodegradation.* J Clin Endocrinol Metab, 1989. **68**(5): p. 882-7.
14. Dalle Carbonare, M. and M.A. Pathak, *Skin photosensitizing agents and the role of reactive oxygen species in photoaging.* J Photochem Photobiol B, 1992. **14**(1-2): p. 105-24.
15. Dissemond, J., et al., *Protective and determining factors for the overall lipid peroxidation in ultraviolet A1-irradiated fibroblasts: in vitro and in vivo investigations.* Br J Dermatol, 2003. **149**(2): p. 341-9.
16. Kahari, V.M. and U. Saarialho-Kere, *Matrix metalloproteinases in skin.* Exp Dermatol, 1997. **6**(5): p. 199-213.
17. Fisher, G.J., et al., *Molecular basis of sun-induced premature skin ageing and retinoid antagonism.* Nature, 1996. **379**(6563): p. 335-9.
18. Lee, K.M., et al., *Analysis of genes responding to ultraviolet B irradiation of HaCaT keratinocytes using a cDNA microarray.* Br J Dermatol, 2005. **152**(1): p. 52-9.
19. Pfeifer, G.P., Y.H. You, and A. Besaratinia, *Mutations induced by ultraviolet light.* Mutat Res, 2005. **571**(1-2): p. 19-31.
20. Cleaver, J.E. and E. Crowley, *UV damage, DNA repair and skin carcinogenesis.* Front Biosci, 2002. **7**: p. d1024-43.
21. McLoone, P., et al., *An action spectrum for the production of cis-urocanic acid in human skin in vivo.* J Invest Dermatol, 2005. **124**(5): p. 1071-4.
22. Van der Molen, R.G., et al., *Broad-spectrum sunscreens offer protection against urocanic acid photoisomerization by artificial ultraviolet radiation in human skin.* J Invest Dermatol, 2000. **115**(3): p. 421-6.
23. Finlay-Jones, J.J. and P.H. Hart, *Photoprotection: sunscreens and the immunomodulatory effects of UV irradiation.* Mutat Res, 1998. **422**(1): p. 155-9.
24. Podda, M., et al., *UV-irradiation depletes antioxidants and causes oxidative damage in a model of human skin.* Free Radic Biol Med, 1998. **24**(1): p. 55-65.
25. Wondrak, G.T., et al., *3-hydroxypyridine chromophores are endogenous sensitizers of photooxidative stress in human skin cells.* J Biol Chem, 2004. **279**(29): p. 30009-20.
26. Choi, C.P., et al., *The effect of narrowband ultraviolet B on the expression of matrix metalloproteinase-1, transforming growth factor-beta1 and type I collagen in human skin fibroblasts.* Clin Exp Dermatol, 2007. **32**(2): p. 180-5.
27. Naru, E., et al., *Functional changes induced by chronic UVA irradiation to cultured human dermal fibroblasts.* Br J Dermatol, 2005. 153 Suppl 2: p. 6-12.
28. Kim, M.S., et al., *Acute exposure of human skin to ultraviolet or infrared radiation or heat stimuli increases mast cell numbers and tryptase expression in human skin in vivo.* Br J Dermatol, 2009. **160**(2): p. 393-402.
29. Sorg, O., et al., *Retinoids in cosmeceuticals.* Dermatol Ther, 2006. **19**(5): p. 289-96.
30. Kang, S., *The mechanism of action of topical retinoids.* Cutis, 2005. **75**(2 Suppl): p. 10-3; discussion 13.
31. Sorg, O., et al., *Oxidative stress-independent depletion of epidermal vitamin A by UVA.* J Invest Dermatol, 2002. **118**(3): p. 513-8.
32. Yaar, M. and B.A. Gilchrest, *Photoageing: mechanism, prevention and therapy.* Br J Dermatol, 2007. **157**(5): p. 874-87.
33. Katsambas, A.D. and A.C. Katoulis, *Topical retinoids in the treatment of aging of the skin.* Adv Exp Med Biol, 1999. **455**: p. 477-82.
34. Van Rijsbergen, J.M. *Opportunities for L(+) lactic acid in the cosmetics industry.* 2004.
35. Hanson, K.M., E. Gratton, and C.J. Bardeen, *Sunscreen enhancement of UV-induced reactive oxygen species in the skin.* Free Radic Biol Med, 2006. **41**(8): p. 1205-12.
36. Pathak, M.A., et al., *Melanin formation in human skin induced by long-wave ultra-violet and visible light.* Nature, 1962. **193**: p. 148-50.
37. Eller, M.S., K. Ostrom, and B.A. Gilchrest, *DNA damage enhances melanogenesis.* Proc Natl Acad Sci U S A, 1996. **93**(3): p. 1087-92.
38. Ortonne, J.P., *Retinoid therapy of pigmentary disorders.* Dermatol Ther, 2006. **19**(5): p. 280-8.
39. Kang, H.Y., et al., *The role of topical retinoids in the treatment of pigmentary disorders: an evidence-based review.* Am J Clin Dermatol, 2009. **10**(4): p. 251-60.
40. Antille, C., et al., *Vitamin A exerts a photoprotective action in skin by absorbing ultraviolet B radiation.* J Invest Dermatol, 2003. **121**(5): p. 1163-7.
41. Varani, J., et al., *Molecular mechanisms of intrinsic skin aging and retinoid-induced repair and reversal.* J Investig Dermatol Symp Proc, 1998. **3**(1): p. 57-60.

# Chapter 06
# THE PREVENTION OF PHOTOAGEING: SUN PROTECTION

One should never underestimate the importance of daily protection from UV rays. Scientists believe that in the past 25 years the ozone layer has been depleted by three per cent. This may not sound like much, but for every one per cent loss of ozone, scientists expect a six per cent increase in skin cancers! In November 1994 the ozone layer over Europe and the USA was 10 to 15 per cent below the normal level. The danger has progressively increased and still the major pollutant-producing countries have not radically changed their production of pollutants. I doubt that we will see any major improvements in the next 30 years because it will take a long time before changes on earth are reflected in the higher atmosphere where the ozone is situated.

Nature tries to deal with excessive sun exposure by two methods: increasing the thickness of the skin and by darkening the colour. A thickened horny layer gives better protection to the skin, but mainly from UVB rays. A tan is the skin's reaction to damaging UV rays. The pigment cells produce more granules of melanin, and melanin absorbs and scatters UV radiation. Melanin is also an extremely powerful free-radical scavenger. Because melanin absorbs all colours of light, it appears black in colour. The absorption of the energy of the whole spectrum of light does not seem to inactivate melanin.

However, melanin is more protective against UVB rays than UVA, so the UVA rays can still damage the deeper layers of the skin. But almost paradoxically, a tan is not a good protection from UV damage; in fact *a tan is a scar*. Although

the melanin produced in the process of tanning becomes protective, the process of producing the tan causes the damage! And remember: this damage is cumulative and does not disappear even if very little exposure follows the damaging episodes.

It is obvious that natural protection is not enough. We are advised to stay out of the midday sun as much as possible but that would condemn us to vitamin D insufficiency which has dire consequences. We should ideally go into the midday sun without sunscreen for an appropriate time according to our skin colour. Then after that apply topical vitamins ACE etc to replace those that have been lost by sun exposure and then put on the sunscreen. Most people will not be able to produce the best levels of vitamin D and for that reason, they should also supplement orally with vitamin D.

## WHAT IS EFFECTIVE, SENSIBLE PROTECTION FROM THE SUN?

For extended periods of sun exposure you should wear sensible clothing made of tightly woven cotton and apply an effective sun protection cream to exposed skin. Remember that we also need to get some UVB in order to make vitamin D, so don't block out the sun too much. Life's like that – a delicate balance of getting just enough and not too much! The safety of sun exposure and the balance of protection are always improved by supplementing the essential vitamins A, C, and E every day.

A person who does not have a clearly sun-damaged skin needs to make Vitamin D by exposing arms, legs, and part of the back and neck for no longer than 20 minutes two or three times per week. A lot has been written about vitamin D deficiency, especially linked to breast cancer and osteoporosis. Many other cancers have now been shown to be linked to vitamin D deficiency, and as we have an ageing population, all of these conditions will become increasingly important.

## WHAT SHOULD YOU LOOK FOR IN A SUNSCREEN?

Real protection from a topically applied sun-protective cream should consist of at least three essential elements to

*"Antioxidants are free radical scavengers that help minimize the damage of UV rays that penetrate the sunscreens and filters."*

be properly effective against excessive, damaging, ultraviolet light. These elements are adequate UVA cover, adequate UVB cover, and sufficient antioxidant cover. Protection elements are either organic molecules, or reflectors, which reflect and scatter the energy, or organic sun filters that chemically absorb part of the sun's UV radiation, or inorganic sunscreens that reflect UV rays.

Sun reflection is undoubtedly more effective and probably much safer than chemical absorption; however, some rays will still manage to penetrate and damage your skin. Because UVA light activates free radicals, it is sensible to use free-radical scavengers at the same time as sun filters or sunscreens. Antioxidants are free radical scavengers that help minimize the damage of UV rays that penetrate the sunscreens and filters.

I believe we should use milder sun protection (like SPF 4 to 8) daily and use the strong creams (about SPF 15 to 20) only when we expose ourselves for protracted periods during sports activities, travel, and outdoor activities. It is important to remember that the SPF rating largely refers to UVB protection, as the index is based on reddening or sunburn, and is therefore an incomplete and possibly misleading index in terms of cancer prevention in skin. Check that you are using both UVA and UVB sunscreens combined with a collection of potent antioxidants.

Scientists do not recommend anything stronger than SPF 15 to 20 because there is minimal advantage from the higher SPFs, but significantly greater doses of sunscreen chemicals. Scientists have recently found that samples of unprotected skin had fewer free radicals in the depth of the skin than skin that had been protected with the ingredients found in the most popular American strong sunscreens. The reason is that stronger creams contain greater concentrations of organic sunscreen agents, which themselves can be converted into free radicals by exposure to UV rays.

A frightening statistic is that skin cancer has risen in the USA and Australia since high SPF products were introduced.[1] We are not sure why. Do people perhaps stay longer in the sun and incur more damage, or are the protective molecules damaging our skin by generating free radicals? It is much better if most or all of the protection is supplied by titanium dioxide and zinc oxide or other reflectant minerals. If you use a purely mineral-based sunscreen without any organic sunscreens, there is no objection to using higher-rated sunscreens. In fact, if one were to use more opaque white sunscreens then virtually 100 per cent of light would be blocked out. Such sunscreens would be very useful to prevent and even control the formation of sun freckles.

## ABSORBENT ORGANIC SUN FILTERS

Absorbent sun filters trap the energy of the UV rays and thereby change their own molecular structure. The molecule

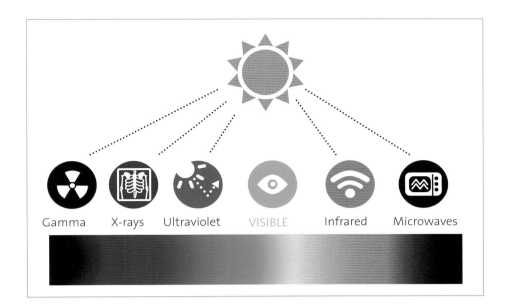

*The electromagnetic spectrum*

is then deactivated as a sunscreen and may be converted into a free radical.

Most filters absorb only the UVB and not the UVA rays, whereas a few may be effective for both UVA and UVB rays. Unfortunately, in many of these cases the effect is usually weaker protection against UVA. On the other hand, there are other chemicals that are partly effective only against UVA and give minimal UVB protection. Newer organic sunscreens are photostable and not permanently deactivated and can be recycled back into a sun-screening molecule and are more effective against UVA.[2] They do not form free radicals on exposure to sunlight.

## UVB PROTECTION

The oldest absorbent sun filter—but today also the most controversial—is para-amino benzoic acid (PABA). PABA is effective in absorbing UVB rays and very effectively prevents reddening of the skin, but paradoxically does not prevent damage of the Langerhans cells (immune cells), which are the most important cells in maintaining healthy skin. PABA has been implicated in facilitating DNA damage and subsequently skin cancer.

Today the most successful UVB protective range is still the methoxycinnamate group that has very little activity in the UVA range. Methoxycinnamates also become free radicals when exposed to UV irradiation, but their protection is greater than the potential free-radical damage if we reapply them every 90 to 120 minutes and use antioxidants at the same time.[3]

Other sunscreens such as octocrylene or benzophenone-3 are mainly UVB sunscreens that also generate free radicals. Benzophenone-34 is often referred to as a broad spectrum UVB and A sunscreen, but it has limited value, and is a potent generator of free radicals. It is difficult to understand why it has not been removed from the list of permitted sunscreen agents.

## UVA PROTECTION

Bemotrizinol (Tinosorb S) and bisoctrizole (Tinosorb M) are effective UVA and UVB sunscreens and are both stable in sunlight and do not become free radicals. Dibenzoylmethane (Parsol 1789) is a most effective UVA screen but, like most molecules, becomes a free radical after absorbing UVA rays. Gradually newer molecules will be introduced that are stable and do not become deactivated or converted into free radicals. Benzophenone is often claimed as a UVA absorber, but it is really a UVB absorber with weak UVA protection. The important point is that we do not know enough about the long-term effects of organic sunscreens. I believe we should rely much more on reflectant inorganic, metallic, or mineral molecules as effective sunscreens.

- Lycopene is a very convenient one because we get it when we eat food like tomatoes (and especially tomato soup or sauce). Parents can ensure that their children are better protected from the sun by giving them lycopene-rich diets, and the protection may last as long as 36 hours. [4, 5, 6]
- Beta-carotene is an important molecule to reduce the UV stress. [7]
- Vitamin C is a way to reduce UV damage. [8]
- Vitamin E reduces sun damage. [9]
- Polypodium leucotomos (a type of tropical fern from the Americas) is another sunscreen agent that can be applied or taken orally to reduce sensitivity to UV damage. [10-12]
- Green tea has wonderful photoprotective properties through the polyphenol molecules and helps minimise photodamage. The richer your skin is in derivatives of green tea, the safer the skin will be in the sun. [23]

Before we leave the subject of organic sunscreens, note that many natural molecules in our skin are in fact powerful organic sunscreens.

The most powerful sun-protective molecule is retinyl palmitate, which in an adequate dose, can have a photoprotective effect equivalent to SPF 20 and has additional protection against UVA. [13] Like virtually every molecule that protects us from sunlight, free radicals may be generated from irradiated retinyl palmitate, but they seem less significant. In skin cell cultures alarm bells have rung because of the possible carcinogenic effects of irradiated retinyl palmitate, [14, 15] but in clinical work, on living humans, exactly the opposite is found. [16-18]

Cluver was the first person to report that oral vitamin A immediately before or after sun exposure reduced the UV damage. [19, 20] At the time he did not consider topical application but, from what we now know, we would achieve photoprotection faster and more efficiently with topical application to specific sites. Oral vitamin A supplementation of up to 75,000 IU per day has been found to be clinically safe and effective in minimizing photodamage and also in reducing the chance of getting skin cancer.[21]

If you regularly nourish yourself with these and numerous other natural molecules, you can be sure that you will be safer when exposed to sunlight and will definitely not interfere with natural vitamin D formation.

## REFLECTANT INORGANIC SUNSCREENS

These agents reflect or block most of the rays (even visible light) and they are opaque, virtually inert, substances. They are more effective in photoprotection than the UV filters, but can cause a white film when used in cosmetic preparations. Most important is titanium oxide, an opaque white powder, ground ultrafine for use in creams. Titanium has the fortunate property of not being recognized as foreign to human tissue, so people who react against other UV filters can use it safely. Titanium dioxide can become a free radical after exposure to sunlight, but only to a very slight degree. [22] It will react with antioxidant vitamins unless specially coated with silica or alumina, and then it is safe to use.

It is important to point out that the 'whitening' of skin after applying a reflective sunscreen is a positive feature from the protection point of view, because it means that an adequate layer of sun-protectant has been placed on the skin. The most effective sun protection, which you can often see on sportsmen, comes from thick pastes made from zinc oxide, which block out sunlight completely. Newer modifications of zinc and titanium dioxide should allow us to make sophisticated transparent sunscreens that also allow inclusion of antioxidant vitamins in the formulations without interacting with the zinc of titanium.

Another very important sun reflectant is zinc oxide, which also has other beneficial effects on skin. It is probably one of the safest molecules that we can apply to our skin and may be the best sun reflectant. It reduces acne and is essential for normal collagen formation. Unfortunately, zinc will react with the antioxidant vitamins.

Remember that there is no truly effective cosmetically acceptable protection from the sun that we can rely on, unless the use of opaque screens comes into fashion! It would be a useful indicator to ensure better coverage of skin if more sunscreens were somewhat opaque. 'If you can't see it, it's not protected'. Transparent products allow no means of judging the quality of cover before sun exposure.

## INFRARED DAMAGE

While a great deal of attention has focused on UVB protection despite the fact that it is only about five per cent of the UV light, almost no one pays any attention to infra-red (IR) light, which is far more prevalent than UV light, and can penetrate deep into the skin. We know IR

*"A supplement of oral Vitamin A of up to 75,000 IU per day has been found to be clinically safe."*

rays more for their heating effects, but relatively little attention has been paid to the damaging effects of the heat on our skin cells. We now recognise that IR can cause epidermal atrophy, pigment blemishes, collagen degeneration, and dermal elastosis—the signs we also see in photoageing. We clearly need protection from IR in our sunscreens. [23, 24]

At this stage we know we have to counter the free radicals induced by IR, so topically applied antioxidants have to be in every sunscreen. IR also induces matrix metalloproteinases[25], which can be minimized by antioxidants.[26] As beta-carotene seems to have a specific role in protecting skin from IR,[27] we should not ignore the importance of taking oral supplements of antioxidants with carotenoids to help minimize IR damage.[28] Hyperforin, an extract of hypericum perforatum, has been shown to be effective at reducing IR damage. [31, 32] More molecules will emerge soon to add to our armamentarium against IR.

The darker the skin is, the more protection it needs from IR, but currently most sunscreens do not take this fact into account.[29] We really need different sunscreens for people with darker skins, who make up over 60 per cent of the world's population! We also need better understanding of all the effects of UV and visible light as well as IR.[30]

## ANTIOXIDANTS AS SUNSCREENS

Antioxidants are a relatively new concept in protecting the skin from damage by UV, visible light, and IR rays, but I have been insisting on using them in sunscreens since 1990. It is important to understand that they do not block UV rays from entering the skin.

How do they work, then? It has been explained in an earlier chapter how light generates free radicals and also that most sunscreens become free radicals when exposed to UV light. If the sunscreen has been absorbed, it aggravates the free radical challenge to the skin, an especially important consideration when treating pigmentation blemishes. Free radicals stimulate the release of active chemicals that induce the melanocyte to make more melanin, so if we are treating pigmentation marks we have to keep the number of free radicals in the skin as low as possible.

Topical antioxidants are attractive as agents to modify this damage because they can mop up excess free radicals. An interesting fact has emerged in my studies: that antioxidants are useful even after sunburn. Apply antioxidants to sunburned skin and you will see that the burned tissue returns to normal at a faster rate than usual. For an even better result add topical vitamin A and also take oral antioxidants and vitamin A.

The most important antioxidants normally found in the skin are: vitamin C, vitamin E, glutathione, alpha-lipoic acid, and coenzyme Q10 (the network antioxidants[33]). Beta-carotene and other carotenoids (such as lutein, lycopene, and zeaxanthine) have an advantage in IR protection. Flavonoids, selenium, superoxide dismutase, and

> *"We really need different sunscreens for people with darker skins, who make up over 60 per cent of the world's population!"*

zinc all assist in giving us better free radical protection.

Protecting the skin cells against the damage caused by ultraviolet light is therefore only a part of logical and sensible protection of the skin. Excess free radical activity is the source of DNA oxidation ('rust') and skin ageing and skin cancer are a function of decay in DNA. Antioxidants are therefore the real protection against damage, especially as it is well known that DNA damage in skin cells, which causes skin cancer, is an incremental, cumulative process. Relying on an SPF factor alone is hopelessly inadequate if one considers the simple facts that explain the mechanisms of damage.

## SUMMARY

We are told that it is wise to protect our skin from the sun, and some people even avoid going into sunlight and consequently suffer from vitamin D deficiency. We have to expose ourselves to sunlight regularly, but we also have to protect ourselves. Sun protection is not a simple topic and the SPF numbers that indicate only the UV protection have deceived us. Increased use of sunscreens has not led to a concomitant reduction in skin cancer. It is better not to rely solely on sunscreen agents but to be sensible about how we expose ourselves to the sun. Our clothing is still the best protection that we have. Our sunscreens should always contain a broad spectrum of antioxidants. Dietary supplements can also go a long way towards reducing UV damage. Protecting the skin cells against the damage induced by ultraviolet light is therefore only a part of protection of the skin. One should remember to reapply sunscreen after swimming or exercise.

Unless sunscreen contains both water- and lipid-soluble antioxidants it should not be considered safe. There is no such thing as a complete sunscreen, regardless of what marketers tell us. Common sense based on a sound understanding of the complex mechanisms involved will provide a balanced approach to sensible sun exposure in light-skinned individuals.

By being aware of what is hype and what is common sense, you can keep yourself safe in the sun and still get all the benefits of sun exposure. I believe that by keeping our skins rich in retinyl palmitate we go a long way towards ideal protection from UV damage while allowing us to make vitamin D naturally.

## REFERENCES

1. Stasko, T., *Is 'Slop' a failure? The current status of photoprotection: a symposium on photoprotection led by Mark Naylor, during the 59th Annual Meeting of the American Academy of Dermatology*, 2001.
2. Gelis, C., et al., *Assessment of the skin photoprotective capacities of an organo-mineral broad-spectrum sunblock on two ex vivo skin models*. Photodermatol Photoimmunol Photomed, 2003. **19**(5): p. 242-53.
3. Hanson, K.M. and R.M. Clegg, *Bioconvertible vitamin antioxidants improve sunscreen photoprotection against UV-induced reactive oxygen species*. J Cosmet Sci, 2003. **54**(6): p. 589-98.
4. Stahl, W. and H. Sies, *Carotenoids and protection against solar UV radiation*. Skin Pharmacol Appl Skin Physiol, 2002. **15**(5): p. 291-6.

5. Cesarini, J.P., et al., *Immediate effects of UV radiation on the skin: modification by an antioxidant complex containing carotenoids*. Photodermatol Photoimmunol Photomed, 2003. **19**(4): p. 182-9.
6. Greul, A.K., et al., *Photoprotection of UV-irradiated human skin: an antioxidative combination of vitamins E and C, carotenoids, selenium and proanthocyanidins*. Skin Pharmacol Appl Skin Physiol, 2002. **15**(5): p. 307-15.
7. Wertz, K., et al., *Beta-carotene interferes with ultraviolet light A-induced gene expression by multiple pathways*. J Invest Dermatol, 2005. **124**(2): p. 428-34.
8. Leveque, N., et al., *Evaluation of a sunscreen photoprotective effect by ascorbic acid assessment in human dermis using microdialysis and gas chromatography mass spectrometry*. Exp Dermatol, 2005. **14**(3): p. 176-81.
9. Foote, J.A., et al., *Chemoprevention of human actinic keratoses by topical DL-alpha-tocopherol*. Cancer Prev Res (Phila Pa), 2009. **2**(4): p. 394-400.
10. González, S. and M.A. Pathak, *Inhibition of ultraviolet-induced formation of reactive oxygen species, lipid peroxidation, erythema and skin photosensitization by polypodium leucotomos*. Photodermatol Photoimmunol Photomed 1996. **12**(2): p. 45-56.
11. Alonso-Lebrero, J.L., et al., *Photoprotective properties of a hydrophilic extract of the fern Polypodium leucotomos on human skin cells*. J Photochem Photobiol B, 2003. **70**(1): p. 31-7.
12. González, S., et al., *Topical or oral administration with an extract of Polypodium leucotomos prevents acute sunburn and psoralen-induced phototoxic reactions as well as depletion of Langerhans cells in human skin*. Photodermatol Photoimmunol Photomed, 1997. **13**(1): p. 50-60.
13. Antille, C., et al., *Vitamin A exerts a photoprotective action in skin by absorbing ultraviolet B radiation*. J Invest Dermatol, 2003. **121**(5): p. 1163-7.
14. Yan, J., et al., *Photo-induced DNA damage and photocytotoxicity of retinyl palmitate and its photodecomposition products*. Toxicol Ind Health, 2005. **21**(7-8): p. 167-75.
15. Tolleson, W.H., et al., *Photodecomposition and phototoxicity of natural retinoids*. Int J Environ Res Public Health, 2005. **2**(1): p. 147-55.
16. Hill, D.L. and C.J. Grubbs, *Retinoids and cancer prevention*. Annu Rev Nutr, 1992. **12**: p. 161-81.
17. Fu, P.P., et al., *Photodecomposition of vitamin A and photobiological implications for the skin*. Photochem Photobiol, 2007. **83**(2): p. 409-24.
18. Hansen, L.A., et al., *Retinoids in chemoprevention and differentiation therapy*. Carcinogenesis, 2000. **21**(7): p. 1271-9.
19. Cluver, E.H., *Sun-trauma prevention*. S Afr Med J, 1964. **38**: p. 801-3.
20. Cluver, E.H. and Politzer, *Sunburn and vitamin A deficiency*. S Afr J Sci, 1965. **61**: p. 306-309.
21. Sedjo, R.L., et al., *Circulating endogenous retinoic acid concentrations among participants enrolled in a randomized placebo-controlled clinical trial of retinyl palmitate*. Cancer Epidemiol Biomarkers Prev, 2004. **13**(11 Pt 1): p. 1687-92.
22. Brezova, V., et al., *Reactive oxygen species produced upon photoexcitation of sunscreens containing titanium dioxide (an EPR study)*. J Photochem Photobiol B, 2005. **79**(2): p. 121-34.
23. Schroeder, P., et al., *Photoprotection beyond ultraviolet radiation--effective sun protection has to include protection against infrared A radiation-induced skin damage*. Skin Pharmacol Physiol, 2010. **23**(1): p. 15-7.
24. Schroeder, P. and J. Krutmann, *What is needed for a sunscreen to provide complete protection*. Skin Therapy Lett, 2010. **15**(4): p. 4-5.
25. Calles, C., et al., *Infrared A radiation influences the skin fibroblast transcriptome: mechanisms and consequences*. J Invest Dermatol, 2010. **130**(6): p. 1524-36.
26. Darvin, M.E., et al., *Radical production by infrared A irradiation in human tissue*. Skin pharmacology and physiology, 2010. **23**(1): p. 40-6.
27. Darvin, M.E., et al., *Topical beta-carotene protects against infra-red-light-induced free radicals*. Experimental dermatology, 2011. **20**(2): p. 125-9.
28. Darvin, M.E., et al., *Determination of the influence of IR radiation on the antioxidative*

network of the human skin. Journal of biophotonics, 2011. **4**(1-2): p. 21-9.
29. Antoniou, C., et al., *Do different ethnic groups need different sun protection?* Skin research and technology : official journal of International Society for Bioengineering and the Skin, 2009. **15**(3): p. 323-9.
30. Dupont, E., J. Gomez, and D. Bilodeau, *Beyond UV radiation: a skin under challenge.* International journal of cosmetic science, 2013. **35**(3): p. 224-32.
31. Arndt, S., et al., *Radical protection in the visible and infrared by a hyperforin-rich cream-in vivo versus ex vivo methods.* Experimental dermatology, 2013. **22**(5): p. 354-7.
32. Meinke, M.C., et al., *In vivo photoprotective and anti-inflammatory effect of hyperforin is associated with high antioxidant activity in vitro and ex vivo.* European journal of pharmaceutics and biopharmaceutics: official journal of Arbeitsgemeinschaft fur Pharmazeutische Verfahrenstechnik e.V, 2012. **81**(2): p. 346-50.
33. Packer, L., G. Rimbach, and F. Virgili, *Antioxidant activity and biologic properties of a procyanidin-rich extract from pine (Pinus maritima) bark, pycnogenol.* Free Radic Biol Med, 1999. **27**(5-6): p. 704-24.

# Chapter 07
# THE CHEMISTRY OF VITAMIN A

*The chemical structure of vitamin A*

### HISTORY

Symptoms, which we now know to be the result of vitamin A deficiency, have been described and treated for centuries. The Ebers Papyrus, written about 1600 BC, probably referred to night blindness when liver was recommended for the eyes, while the Chinese in 1500 BC were prescribing liver and honey as a cure for night blindness. Hippocrates advised the regular consumption of the whole liver of an ox dipped in honey.[39] For more than 3000 years liver, one of the richest sources of vitamin A (mainly as retinyl palmitate) has been used to treat the earliest sign of vitamin A deficiency.

Although it has been known since ancient Egyptian times that certain foods, such as liver, would cure night blindness, vitamin A per se was not identified until 1912 by Frederick Gowland Hopkins, who received a Nobel prize for his work on 'accessory food factors' which later became known as vitamins. Vitamin A's chemical structure was defined in 1931 by Paul Karrer, who also received a Nobel Prize in 1937 for his work, because this was the first time that a vitamin's structure was determined.

Vitamin A belongs to a family of chemicals called 'retinoids', so called because of their close association with the chemistry of vision in the retina of the eye. Vitamin A is essential for the production of visual purple. By the mid 1930s they had defined the chemistry of the vitamin A molecule, and by 1947 they were able to synthesised vitamin A as retinoic acid and retinyl palmitate in the laboratory.

There are four main forms of vitamin A but they all have essentially the same action:

**I. RETINOL:** this is generally accepted as the basic form of vitamin A from a chemical point of view. Retinol as the alcohol form is a very active form of vitamin A and is used to transport the

> "Vitamin A belongs to a family of chemicals called 'retinoids', so called because of their close association with the chemistry of vision in the retina of the eye."

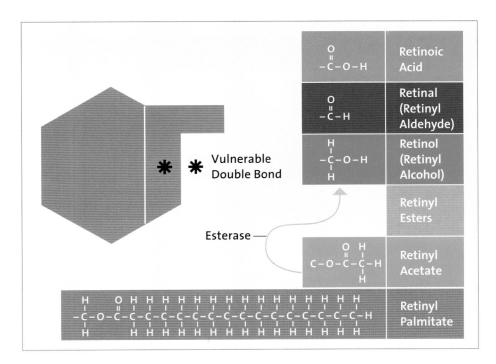

*The Chemistry of vitamin A*

# HERE IS A VITAMIN A AND CAROTENOID TIME LINE:

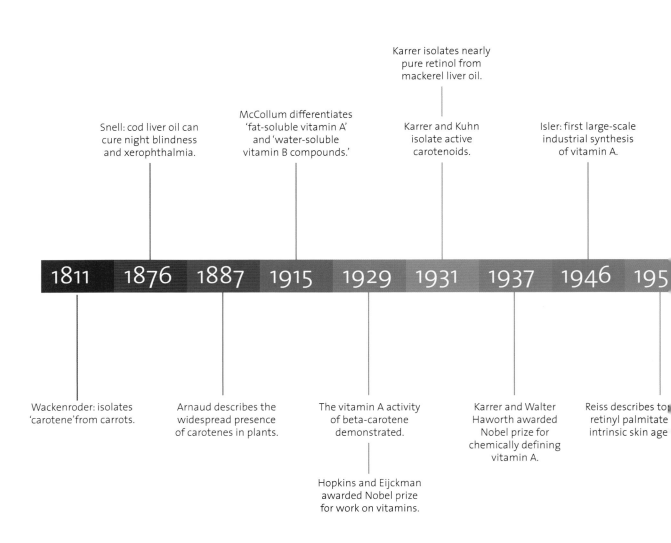

66 \ The Chemistry of Vitamin A

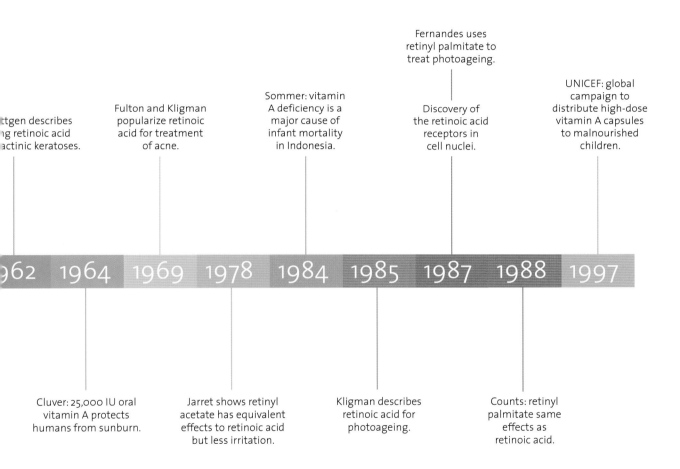

The Chemistry of Vitamin A / 67

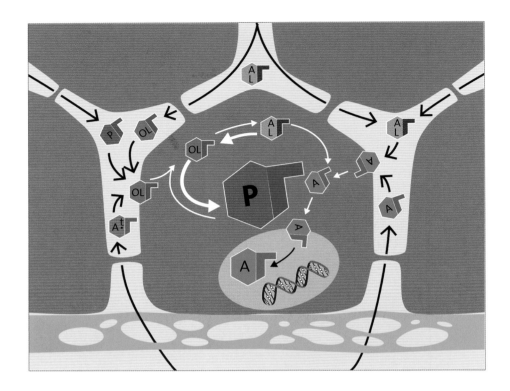

*Diagram of a basal keratinocyte being nourished with either retinyl palmitate, retinol, retinaldehyde or retinoic acid which enters the interstitial fluid and then is taken up by retinoid receptors into the cytosol of the cell and, virtually immediately, converted into retinyl palmitate and other esters which are gradually metabolized through retinol, to retinal and then to retinoic acid. Retinoic acid cannot be stored as retinyl esters and can only act on the nuclear receptors and then enter the nucleus and act on the retinoid responsive genes.*

vitamin in the bloodstream. It is less irritating than retinoic acid.

The main difficulty with retinol lies in its chemical instability, making it difficult to formulate a cream and keep it active. Like the other forms of vitamin A, it is sensitive to light, air (oxygen), and water. Retinol is less sensitive to light, especially when it is joined to the proteins that transport it, and in cells it is less sensitive to light than retinyl palmitate.[6] In general, retinol has a limited shelf life and is best protected in liposomes.

Retinol is less irritating to skin than retinoic acid but is much more irritating than retinyl acetate or retinyl palmitate. Retinol is normally found only in extremely low doses in tissues, and when higher doses are given to cells, the cell membranes may be damaged. This may explain in part why retinol often causes mild peeling of skin when it is first used. As the surface of the cell walls develop more retinoid receptors, the irritation disappears. Retinol is otherwise just another vitamin A and will produce the same results if used in the same international unit dose as retinyl acetate.

Retinol can pass the placental barrier and enter the bloodstream of the developing foetus [7] to supply it with vitamin A, an essential nutrient for normal foetal development. Retinol is two metabolic steps away from retinoic acid, and that fact deceives many clinicians into believing that it will deliver retinoic acid more efficiently than retinyl palmitate. What seems to be ignored is that retinol is only one metabolic step away from retinyl palmitate, and that is the preferred metabolic route for retinol, especially when applied to the skin. [8] Topically applied retinol is almost completely converted to retinyl palmitate and only a tiny fraction remains as retinol that can then be metabolized to retinaldehyde and then to retinoic acid.

One may ask why not use retinol in some skin care products if it is going to be converted into retinyl palmitate. Various isomers (forms) of vitamin A acid are essential to give the full effects of vitamin A. By supplying vitamin A in its various forms, one hopes to increase the chances of a widespread mixture of retinoic acid isomers. Furthermore, retinol can be ionized easily and is useful with iontophoresis because it can be transported deeper into the skin by ion flow.

**II. RETINALDEHYDE (RETINAL)** is the form of vitamin A we use to make visual purple, which is essential for night vision. It has also been used topically on skin because it is only one metabolic step away from retinoic acid.[9, 10] Retinol is changed by oxidation into retinaldehyde, and retinaldehyde in turn is oxidized one step further into retinoic acid. Retinaldehyde can make the same changes as retinoic acid[9] because it is metabolized into retinoic acid.[11] When retinal is applied to the skin, however, virtually all of it is converted into retinyl esters and only a tiny fraction is converted to retinoic acid.[12, 13]

**III. RETINOIC ACID (RETIN A)** is the metabolically active form of vitamin A, which works on the DNA of the cell nucleus. Retinoic acid is generally only available on prescription for topical use. While we talk about retinoic acid as if there were only one form, in fact there are a number of isomers (variants with a different spatial arrangement of the atoms in the molecule), and each isomer has special functions, which we can briefly explore. All-trans-retinoic acid (ATRA), 9-cis-retinoic acid, all-trans-3, 4-didehydroretinoic acid8, and 14-hydroxy-4, 14-retroretinol are the ligands of retinoic acid. These specifically bind – either alone or with each other or with other hormones like vitamin D, thyroid hormone, testosterone – with the RAR and RXR nuclear receptors to produce all the amazing effects of vitamin A. Unfortunately, we could never describe all this in chemical detail in a text book. The chemistry is ever burgeoning because vitamin A plays such a substantial role in the normal cellular activity of both stem cells and highly differentiated cells. While these details are fascinating, we do not need to understand the intricacies in order to use vitamin A sensibly in skin care.

**IV. RETINYL PALMITATE,** retinyl propionate, and retinyl acetate are chemically more stable forms of vitamin A that are milder, but still active, and

*If retinol is applied to the surface of the skin it is taken up into the keratiinocytes and more than 90 percent is converted into retinyl esters. Some retinol may be metabolized to retinal and then retinoic acid.*

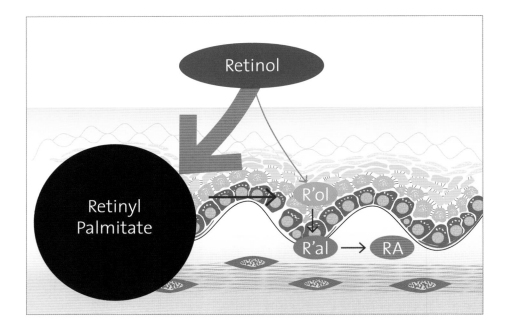

easier for the skin to tolerate. Retinyl esters and particularly retinyl palmitate are the dominant forms in which vitamin A is stored in the liver, in the skin, and in cells all over the body. Vitamin A is necessary for the formation of healthy blood cells in the bone marrow as well as for making healthy bones. In fact, vitamin A is essential for every cell of the body, but because those cells are protected from light irradiation, they do not become deficient of vitamin A. More than 80 per cent of the vitamin A normally found in the skin is in the form of retinyl palmitate.[13] Retinyl acetate is more active than palmitate and about as active as retinoic acid when applied to the skin.[14]

There is a natural interplay between the various forms of vitamin A in the skin. Generally about 91 per cent of all vitamin A in the skin is in the form of retinyl esters (mainly retinyl palmitate), and these are metabolized reversibly to retinol, which constitutes about 3 per cent of the vitamin A.

Retinol in turn can be reversibly converted to retinaldehyde that constitutes another 3 per cent of the total vitamin A. Retinoic acid is also found physiologically at about 3 per cent of the total vitamin A content of the skin.[15, 16] This amount is very tightly controlled because some of the transient forms (retinol, retinoic acid) are cellular irritants. In normal circumstances free retinol is hardly ever found in the interstitial fluid, while retinoic acid is found almost exclusively in the area of the nucleus, for that is where it works. If there is an excess of retinol or retinaldehyde, it is preferentially converted via retinol into retinyl esters. Only a small amount of retinaldehyde is converted irreversibly to retinoic acid that is then used in the nucleus of the cells to interact with the DNA.

Retinoic acid cannot be stored, whereas all the other forms of vitamin A can be stored as retinyl esters. It is easy to load the skin with topical applications of retinol, retinal, or retinyl esters.[17] In general, all physiological effects of retinoic acid can be achieved by applying any form of vitamin A provided that the dose applied is relatively equivalent to the dose of retinoic acid that is required to have an effect.

## THE CONDITIONS NECESSARY FOR FORMULATING VITAMIN A PRODUCTS

Retinoids are all very active chemicals, easily degraded by exposure to light,[18] heat, air, and moisture. Therefore, when used in cosmetics they must be made in a strictly controlled environment, preferably under red light to prevent photo-degradation of vitamin A during manufacture. Afterwards the preparations must be carefully protected in their containers and not put into conventional cosmetic glass jars with wide lids, because the vitamin A on the surface of the cream would be denatured by exposure to air and light. As a result, any vitamin A product in a glass or plastic jar will not actually deliver the intended dose of vitamin A.

All creams using retinyl palmitate, retinyl acetate or retinol creams/gels have to be packed in collapsible aluminium laminated tubes or airless dispensers that protect these important vitamins from light and allow very little air to be sucked back into the container to keep the vitamins as active and stable for as long as possible.

## VITAMIN A AND BETA-CAROTENE

In plants we do not find vitamin A but instead carotenoids like beta-carotene, which is composed of two molecules of the form of vitamin A used for night vision (retinyl aldehyde). We eat beta-carotene and other carotenoids by including such major sources as spinach, parsley, carrots, tomatoes, and broccoli in our diet. All the vitamin A found in vegetarian animals (like cattle and sheep) comes from carotenoids.

The real significance of vitamin A in skin ageing was realized by two research workers in the 1930s[19]. They knew that vitamin A was very easily degraded by exposure to light and air and postulated that because old skin was generally sun-exposed skin, the cause of aged skin could be the result of a localized deficiency of vitamin A caused by sunlight destroying the vitamin A molecule. The unravelling of this complex story started in South Africa as long ago as the 1950s when Professor Cluver proved that every time we go out into sunlight we significantly deplete our vitamin A, not only in the skin but also in the blood.[20] He also showed that in winter the blood levels

*"Professor Cluver proved that every time we go out into sunlight we significantly deplete our vitamin A"*

of vitamin A are lower than in summer. He advised people to use vitamin A immediately before going into the sun and also after heavy exposure to sunlight to prevent severe sunburn. The dose he recommended was about 25,000 IU per tablet. It was quite clear that the vitamin A was acting as one of the very first, and most natural, sunscreens. If high doses of vitamin A (as retinyl palmitate/acetate at about 40,000 IU) are given orally, then the acute effects of sunburn can be minimized and even reversed!

UVA destroys vitamin A by altering the molecule in a subtle way. Once this change occurs, it cannot be reversed, and that molecule no longer has any vitamin A activity. When you realize that UVA can penetrate through clouds, through window glass, and even through many clothing materials down into your dermis, then you get some idea of the really serious problem that we face.

The vitamin A in our skin seems destined to be destroyed each time we go out into the sun, but vitamin A is acting as a natural sunscreen and the richer the skin is in vitamin A, the more protected it will be against UV damage. However, in ordinary circumstances without vitamin A supplementation orally or topically, this exposure causes a localized deficiency that would not be important if it were not for the fact that we go out into daylight every day. Our skin slowly starts to suffer from this deficiency soon after our first exposure to daylight as a child. This becomes a chronic low-grade deficiency because the body cannot manage to resupply the skin as rapidly as the vitamin A is destroyed. It has been estimated that if you spend a day at the beach and get a good tan, then you have probably destroyed more than 90 per cent of the vitamin A in your skin.

If you then stay in a dark room and wait for the skin levels of vitamin A to be restored to normal, you are going to spend about five to seven days in that dark room.[20, 21] Of course, you could speed things up by applying the vitamin A directly to the skin.[22] More about this later.

## THE ACTION OF RETINYL PALMITATE AND RETINYL ACETATE

Retinyl palmitate was the first form of vitamin A proved to be effective in rejuvenating old skin. This research work was done in 1955 but no one seems to have taken much note of the article.[1] Presumably people tried to use it for only a short while and were discouraged by the very slow changes induced. Significant research has been conducted in Europe and the USA on the activity of the various forms of vitamin A. Recent research in the UK has confirmed that retinyl palmitate will give the same results as retinoic acid.[23]

Retinyl palmitate/acetate enters cells and is stored as retinyl esters (mainly retinyl palmitate) and converted as required by normal physiology to retinol, retinaldehyde, and finally to retinoic acid. Retinoic acid accounts for virtually all of the effects of vitamin A.[15] The palmitate moiety of the molecule has a very important role because it seems a highly desirable molecule for cells and is easily absorbed.

## EFFECTS OF RETINYL PALMITATE AND RETINYL ACETATE

Cell walls are not easy to penetrate. Mechanisms have to exist to facilitate the transfer of a chemical from outside the cell wall through the membrane into the cell substance itself. It has been discovered that there are little secret passageways through the cell wall for various chemicals. Only the chemicals with a required 'password' or shape may enter any particular passageway. There are places on the keratinocyte cell wall, 'cellular retinoid receptors', where vitamin A as retinyl palmitate, acetate, retinol, or retinoic acid can enter the cell. Once inside the cell itself, the vitamin A is stored as retinyl esters and slowly released to form retinol, then retinaldehyde, and finally retinoic acid in its various forms. Probably only retinoic acid enters the central nucleus and acts on the DNA to produce the following changes:

❶ Retinoic acid affects about 300 to 1000 genes of the human body [24] and in particular the genes that cause epidermal stem cells to grow and differentiate into normal keratinocytes and mature into healthy layers of the epidermis. [25] It is basically a regulator or normalizer of DNA. It increases the growth of the basal layer (growth layer) of skin cells, which is why the epidermis becomes thicker. Although not confirmed specifically, this effect is almost certainly mediated by the induction of a complex release of growth factors in their ideal relative concentrations.

❷ Not only does the skin get thicker [26], it also heals faster, because the cells multiply at a faster rate and the keratinocyte cycle from basement up to the shedding from the stratum corneum is speeded up. Our skin cells are generally deficient of vitamin A and grow slower as we age. Vitamin A does not make the cells grow faster than they normally would, it just restores the skin growth to normal. You need not worry that you will speed up the

*Clinical changes after eight months of using low dosage vitamin A with added vitamins C, E, B5 and carotenoids. The patient's skin is less inflamed, smoother and well hydrated.*

cellular growth to a degree that will evoke the so-called Hayflick limit and shorten the active life of your skin. (The Hayflick limit is the proposed number of times a cell will divide before it stops dividing because of systematic loss of genetic material with every division.) Your skin will just be healthier.

❸ Topical vitamin A helps skin to withstand environmental damage.[27] Retinyl palmitate specifically (and also retinyl esters to a similar degree) acts as a powerful natural sunscreen by absorbing UV rays and protecting DNA from damage.[28] This property is confined to the esters and is quite different from retinol, retinal, and retinoic acid, which are all photosensitizers and aggravate the damage from sunlight.

❹ If the skin is kept sufficiently rich in vitamin A, we prevent the signs of photodamage even when we go out into the sun. The levels of vitamin A are not significantly depleted, and the skin remains protected from UV damage. The important fact is that the effects of retinoic acid on DNA are maintained. One of the benefits is that fewer matrix-metalloproteinases (MMPs) are released with less damage to the collagen.[29] There is also less damage to the DNA of keratinocytes, and consequently melanin production is not upgraded.[30, 31] This benefit is entirely dose-related, and from clinical experience it will generally only occur with high levels of cutaneous vitamin A.

❺ Sun damage flattens dermal papillae, and topical vitamin A helps to restore them and causes the epidermis to be better nourished.[32] This improvement manifests as greater stability of the epidermis on the dermis and consequently less fine wrinkling of the skin.

❻ Vitamin A affects the fibroblast cells, the most important cell in the dermis, and particularly induces the genes for the production of collagen.[33] Healthier lattice-type collagen I and collagen III are formed, and unhealthy collagen is removed by enzyme activity. It is likely that this is due to stimulation of TGF-beta[34] and particularly TGF-beta-3. Maybe as research targets this area specifically, we will get more clarification.

❼ Vitamin A as retinyl esters is essential for preventing the development of skin cancers.[35-37] People with skin cancers have been found to have low levels of retinyl esters in their skin.[38] By raising the vitamin A levels and keeping high levels in the skin throughout life, we can make it unlikely that cancer will develop in the skin.[39] Vitamin A also reduces the actions of ornithine decarboxylase, which independently reduces the chances of developing skin cancer.[40]

⑧ Vitamin A increases the secretion of glycosaminoglycans (GAGS, natural moisturising factors) by the fibroblast cells into the space between the cells, allowing the skin to retain more water with some puffing out of the wrinkles.[41] Increased hyaluronic acid also occurs[42, 27] particularly in sun-damaged skin.[43] These natural moisturizing factors also filter up into the epidermis between the cells. GAGs constitute most of the chemicals created by the fibroblast to help retain moisture.

⑨ Vitamin A improves the quality of elastin fibres, though it seems not to increase the quantity of elastin but rather seems to remove damaged elastin.[44] The elastin is most probably removed by white blood cells that clean up healthy skin.

⑩ The blood supply to the deeper layers of the skin is improved, which means that more natural foods and oxygen are delivered to the skin.[32]

Retinyl palmitate and retinyl acetate repair ultraviolet damage by virtue of the fact that they are converted to retinoic acid in the cell. As a result they also prevent tissue wasting and the destruction of collagen, which is generally found with ageing. They will do everything that retinoic acid can do, because ultimately they are converted into retinoic acid provided that they are used in the same number of international units as the retinoic acid.

## CONCENTRATIONS OF VITAMIN A

For convenience vitamin A is measured in international units (IU) per gram (or in the food industry they use retinol equivalents: R.E., which are 3.3 IU). The recommended effective doses lie between 500 IU and 50,000 IU. Anything less than 500 IU is generally of minimal value unless used with some penetrant enhancer.

One can achieve effective vitamin-A-based products only if one takes into account the importance of the international unit measurement. Ten IU of retinol produce certain consistently measurable biological results. Retinyl palmitate 10 IU will produce exactly the same biological result as retinol 10 IU but will weigh considerably more than retinol. The same is true for retinoic acid, but it weighs even less than retinol. Therefore the international unit measurement ensures comparable results and is far more accurate than percentages.

Just a fraction of 1 per cent of retinoic acid can be much more powerful than 2 per cent of retinol, but 1000 IU retinoic acid will give exactly the same effects as 1000 IU retinol despite a vast difference in their weight and percentage rating in the formula. A more recent unit of measurement is the Retinol Activity Equivalent (RAE), which is an attempt to define more accurately the comparative activity of the different vitamin A molecules. One RAE will produce the same result no matter which chemical is used.

The safety margin of retinyl palmitate and retinyl acetate is enormous when

*"Incidentally, high levels of vitamin A in the skin cannot produce systemic effects like liver damage, interference with foetal development, or other complications of too much vitamin A."*

> *"The foetus needs healthy doses of vitamin A in order to avoid foetal abnormalities!"*

applied in recommended doses to the skin. Topical application is usually without reaction at high dose provided that the skin has been primed by lower doses to initiate the production of cellular retinoid receptors.

Retinoid receptors are very susceptible to UV light and in sun-exposed skin there are lower than normal levels of retinoid receptors, and as a result the cell cannot absorb vitamin A. The vitamin A in the extracellular fluid acts as a chemical irritant which causes a transient retinoid reaction. The apparent paradox is that whatever vitamin A is absorbed helps to build more retinoid receptors, and so more vitamin A can be absorbed and the retinoid reaction slowly disappears.

Incidentally, high levels of vitamin A in the skin cannot produce systemic effects like liver damage, interference with foetal development, or other complications of too much vitamin A. This is because the vitamin A is virtually locked in the skin[45] due to the fact that the enzymes required to mobilize it by combining it with a lipoprotein do not exist in the skin. As a result, topically applied vitamin A products do not cause any changes to serum levels of vitamin A.[46]

Naturally occurring forms of vitamin A such as retinol and its active metabolites are essential for vision and control the epithelial cells of the gastro-intestinal tract, respiratory system, skin, and bone. In addition they are vital for the nervous system and the immune system and for hematopoiesis. Retinoids are essential for growth, reproduction (conception and embryonic development), and resistance to, and recovery from, infection. Vitamin A starts work soon after conception and continues throughout our whole lifespan.

Many women worry about using topical vitamin A products while pregnant because popular myth has implicated vitamin A in birth defects. People who spread these rumours are probably unaware that they are incorrectly extrapolating from our experience with the oral use of cis-retinoic acid for acne. There is no doubt that cis-retinoic acid has caused birth defects, apparently because of high doses. The average dose of cis-retinoic acid is about 300,000 IU per day. Research in Sweden suggests that 'repeated oral doses of up to 30,000 IU of vitamin A in addition to dietary vitamin A are without safety concern. Safe doses are probably higher, since plasma concentrations and exposure to RA remained at levels earlier shown to be without increased risk of teratogenicity in pregnant women'.[47] People forget that a growing foetus is the most perfect example of cells that are growing, differentiating and maturing – exactly the areas where vitamin A exerts such important effects in all cells of the body. The foetus needs healthy doses of vitamin A in order to avoid foetal abnormalities! Despite general perceptions about vitamin A that are not supported by scientific research, we can reassure you that vitamin A in healthy oral doses is not dangerous. As for topical application of vitamin A, you can be sure that it is safe to use at any dose because it cannot get into the bloodstream.

Now that we know that Vitamin A is safe and even essential for normal healthy skin, let's see in the next chapter how we use topical vitamin A to make healthy skin.

## REFERENCES

1. Reiss, F. and R.M. Campbell, *The effect of topical application of vitamin A with special reference to the senile skin*. Dermatologica, 1954. **108**(2): p. 121-8.
2. Stuttgen, G., *Zur Lokalbehandlung von Keratosen mit Vitamin A-Säure* . Dermatologica, 1962. **124**: p. 65-80.
3. Cluver, E.H., *Sun-trauma prevention*. S Afr Med J, 1964. **38**: p. 801-3.
4. Sommer, A., *Large dose vitamin A to control vitamin A deficiency*. International journal for vitamin and nutrition research. Supplement = Internationale Zeitschrift fur Vitamin- und Ernahrungsforschung. Supplement, 1989. **30**: p. 37-41.
5. Debska, O., G. Kaminska-Winciorek, and R. Spiewak, *[Does sunscreen use influence the level of vitamin D in the body?]*. Polski merkuriusz lekarski : organ Polskiego Towarzystwa Lekarskiego, 2013. **34**(204): p. 368-70.
6. Tang, G., et al., *Epidermis and serum protect retinol but not retinyl esters from sunlight-induced photodegradation*. Photodermatol Photoimmunol Photomed, 1994. **10**(1): p. 1-7.
7. Biesalski, H.K., et al., *Long-term administration of high dose vitamin A to rats does not cause fetal malformations: macroscopic, skeletal and physicochemical findings*. J Nutr, 1996. **126**(4): p. 973-83.
8. Duell, E.A., et al., *Extraction of human epidermis treated with retinol yields retro-retinoids in addition to free retinol and retinyl esters*. J Invest Dermatol, 1996. **107**(2): p. 178-82.
9. Saurat, J.H., et al., *Topical retinaldehyde on human skin: biologic effects and tolerance*. J Invest Dermatol, 1994. **103**(6): p. 770.
10. Didierjean, L., et al., *Biological activities of topical retinaldehyde*. Dermatology, 1999. **199 Suppl 1**: p. 19-24.
11. Didierjean, L., et al., *Topical retinaldehyde increases skin content of retinoic acid and exerts biologic activity in mouse skin*. J Invest Dermatol, 1996. **107**(5): p. 714-9.
12. Sorg, O., L. Didierjean, and J.H. Saurat, *Metabolism of topical retinaldehyde*. Dermatology, 1999. **199 Suppl 1**: p. 13-7.
13. Vahlquist, A., *Vitamin A in human skin: I. detection and identification of retinoids in normal epidermis*. J Invest Dermatol, 1982. **79**(2): p. 89-93.
14. Jarrett, A., R. Wrench, and B. Mahmoud, *The effects of retinyl acetate on epidermal proliferation and differentiation. I. Induced enzyme reactions in the epidermis*. Clin Exp Dermatol, 1978. **3**(2): p. 173-88.
15. Kurlandsky, S.B., et al., *Auto-regulation of retinoic acid biosynthesis through regulation of retinol esterification in human keratinocytes*. J Biol Chem, 1996. **271**(26): p. 15346-52.
16. Yan, J., et al., *Levels of retinyl palmitate and retinol in the stratum corneum, epidermis, and dermis of female SKH-1 mice topically treated with retinyl palmitate*. Toxicol Ind Health, 2006. **22**(4): p. 181-91.
17. Sorg, O., C. Tran, and J.H. Saurat, *Cutaneous vitamins A and E in the context of ultraviolet- or chemically-induced oxidative stress*. Skin Pharmacol Appl Skin Physiol, 2001. **14**(6): p. 363-72.
18. Berne, B., M. Nilsson, and A. Vahlquist, *UV irradiation and cutaneous vitamin A: an experimental study in rabbit and human skin*. J Invest Dermatol, 1984. **83**(6): p. 401-4.
19. Sulzberger, M.B. and F. Wise, *The year book of dermatology and syphilology*. 1938 ed1938, Chicago: Year Book Publishers
20. Cluver, E.H. and W.M. Politzer, *The pathology of sun trauma*. S Afr Med J, 1965. **39**(41): p. 1051-3.
21. Cluver, E.H. and Politzer, *Sunburn and vitamin A deficiency*. S Afr J Sci, 1965. **61**: p. 306-309.
22. Spearman, R.I. and A. Jarrett, *Biological comparison of isomers and chemical forms of vitamin A (retinol)*. Br J Dermatol, 1974. **90**(5): p. 553-60.
23. Watson, R.E., et al., *Repair of photoaged dermal matrix by topical application of a cosmetic 'anti-ageing' product*. Br J Dermatol, 2008. **158**(3): p. 472-7.
24. Fadloun, A., et al., *Retinoic acid induces TGFbeta-dependent autocrine fibroblast growth*. Oncogene, 2008. **27**(4): p. 477-89.

25. Satish, L., et al., *Gene expression patterns in isolated keloid fibroblasts.* Wound Repair Regen, 2006. **14**(4): p. 463-70.
26. Bryce, G.F. and S.S. Shapiro, *Retinoid effects on photodamaged skin.* Methods Enzymol, 1990. **190**: p. 352-60.
27. Goffin, V., et al., *Topical retinol and the stratum corneum response to an environmental threat.* Skin Pharmacol, 1997. **10**(2): p. 85-9.
28. Antille, C., et al., *Vitamin A exerts a photoprotective action in skin by absorbing ultraviolet B radiation.* J Invest Dermatol, 2003. **121**(5): p. 1163-7.
29. Watson, R.E., et al., *Retinoic acid receptor alpha expression and cutaneous ageing.* Mech Ageing Dev, 2004. **125**(7): p. 465-73.
30. Sato, K., et al., *Depigmenting mechanisms of all-trans retinoic acid and retinol on B16 melanoma cells.* Biosci Biotechnol Biochem, 2008. **72**(10): p. 2589-97.
31. Ortonne, J.P., *Retinoid therapy of pigmentary disorders.* Dermatol Ther, 2006. **19**(5): p. 280-8.
32. Gilchrest, B.A., *A review of skin ageing and its medical therapy.* Br J Dermatol, 1996. **135**(6): p. 867-75.
33. Kang, S., *The mechanism of action of topical retinoids.* Cutis, 2005. **75**(2 Suppl): p. 10-3; discussion 13.
34. Choi, Y. and E. Fuchs, *TGF-beta and retinoic acid: regulators of growth and modifiers of differentiation in human epidermal cells.* Cell regulation, 1990. **1**(11): p. 791-809.
35. Yaar, M. and B.A. Gilchrest, *Photoageing: mechanism, prevention and therapy.* Br J Dermatol, 2007. **157**(5): p. 874-87.
36. Guo, X. and L.J. Gudas, *Metabolism of all-trans-retinol in normal human cell strains and squamous cell carcinoma (SCC) lines from the oral cavity and skin: reduced esterification of retinol in SCC lines.* Cancer Res, 1998. **58**(1): p. 166-76.
37. Saurat, J.H., *Skin, sun, and vitamin A: from ageing to cancer.* J Dermatol, 2001. **28**(11): p. 595-8.
38. Guo, X., et al., *Esterification of all-trans-retinol in normal human epithelial cell strains and carcinoma lines from oral cavity, skin and breast: reduced expression of lecithin: retinol acyltransferase in carcinoma lines.* Carcinogenesis, 2000. **21**(11): p. 1925-33.
39. Sorg, O., et al., *Retinol and retinyl ester epidermal pools are not identically sensitive to UVB irradiation and anti-oxidant protective effect.* Dermatology, 1999. **199**(4): p. 302-7.
40. Dawson, M.I., et al., *Retinoic acid (RA) receptor transcriptional activation correlates with inhibition of 12-O-tetradecanoylphorbol-13-acetate-induced ornithine decarboxylase (ODC) activity by retinoids: a potential role for trans-RA-induced ZBP-89 in ODC inhibition.* Int J Cancer, 2001. **91**(1): p. 8-21.
41. Kafi, R., et al., *Improvement of naturally aged skin with vitamin A (retinol).* Arch Dermatol, 2007. **143**(5): p. 606-12.
42. Sayo, T., S. Sakai, and S. Inoue, *Synergistic effect of N-acetylglucosamine and retinoids on hyaluronan production in human keratinocytes.* Skin Pharmacol Physiol, 2004. **17**(2): p. 77-83.
43. Calikoglu, E., et al., *UVA and UVB decrease the expression of CD44 and hyaluronate in mouse epidermis, which is counteracted by topical retinoids.* Photochemistry and Photobiology, 2006. **82**(5): p. 1342-7.
44. El-Domyati, M., et al., *Intrinsic ageing vs. photoageing: a comparative histopathological, immunohistochemical, and ultrastructural study of skin.* Exp Dermatol, 2002. **11**(5): p. 398-405.
45. Sass, J.O., et al., *Plasma retinoids after topical use of retinaldehyde on human skin.* Skin Pharmacol, 1996. **9**(5): p. 322-6.
46. Nohynek, G.J., et al., *Repeated topical treatment, in contrast to single oral doses, with Vitamin A-containing preparations does not affect plasma concentrations of retinol, retinyl esters or retinoic acids in female subjects of child-bearing age.* Toxicol Lett, 2006. **163**(1): p. 65-76.
47. Hartmann, S., et al., *Exposure to retinyl esters, retinol, and retinoic acids in non-pregnant women following increasing single and repeated oral doses of vitamin A.* Ann Nutr Metab, 2005. **49**(3): p. 155-64.

# Chapter 08
# THE ROLE OF VITAMIN A IN PHOTOAGEING

We have discussed the prevention of photoageing as mainly protection from sun damage; however, that is only one part of the story. I pointed out in the previous chapter that one can minimize photoageing in many other ways besides sunscreens. The one item that stands out above all others is the role of vitamin A, which has such a central role in skin health, protection from photodamage, and treatment of photoageing.

At the beginning of the twentieth century researchers started looking for the essential chemicals that cannot be made in our bodies and have to be supplied by the diet to maintain healthy cellular functions. They knew that a deficiency of these chemicals caused disease, and because they thought that these substances were amines (a special group of chemicals) and were essential for life, they were called 'vital amines', which became 'vitamins'.

Vitamin A has been known since 1911 as a fat-soluble chemical found in fish liver oils, egg yolk, milk, and animal livers, and it soon became known as the skin vitamin. It is unstable in light because it absorbs the energy of UV rays, and that is why it has become especially vulnerable in skin. It is a paradox that the most essential vitamin in the skin is damaged by the main feature of the environment that the skin has to encounter! However, let's look at it another way: the reason

*"Vitamin A has been known since 1911 as a fat-soluble chemical found in fish liver oils, egg yolk, milk, and animal livers"*

it is destroyed is that it acts as a very powerful natural sunscreen.

There is no other molecule that has a similar role, and because vitamin A plays such a vital role in our metabolic processes and DNA activity, it will for all time be the fundamental essential molecule to keep skin healthy and rejuvenate old skin. There is no alternative to it because this is the vitamin that works intimately and naturally with our DNA to determine how stem cells behave, how cells differentiate into specific cells, and how they mature into fully functional healthy cells, not only in our skin but also throughout the body. Vitamin A is one of the most misunderstood nutrients in the world of cosmetics, and yet it is the most fascinating compound that will do almost everything one needs to keep skin healthy and radiant.

Some people develop a 'sun phobia' and avoid sunlight as much as possible in order to avoid the chances of wrinkles, melasma, and skin cancer. This raises an important dilemma: are the dangers of sun exposure greater or less than the dangers of avoiding sunlight? If you avoid sunlight altogether, especially in summer, then you cannot make natural vitamin D, and that will increase the chances of osteoporosis, [1] which is very common in Japan, for example.[2] Vitamin D protects us from UV rays[3] and also protects us from various cancers.[4-5]

The question is, therefore: would you prefer to get a skin cancer or a breast, bowel, or prostatic cancer? Or in one of the greater ironies, to increase the risk of developing malignant melanoma as a result of too little vitamin D? Is pale smooth skin alone a good exchange for the risks of vitamin D deficiency? I think not. I believe we have to find a way to get all the benefits of sunlight and minimize the dangers. This is where retinyl palmitate (RP) can help us significantly. Moreover, retinoids share receptors with vitamin D and together they are responsible for the delicate balance required for normal skin.[6]

## VITAMIN A METABOLISM

The normal metabolism of vitamin A is as follows: About 80 to 90 per cent of all vitamin A in a cell is RP (retinyl palmitate) or similar esters of vitamin A. RP is de-esterified to retinol and then changed to retinal and finally converted to retinoic acid[7]. There seems to be a relatively fixed ratio of approximately 90 per cent RP to 3 per cent retinol, 3 per cent retinal and 3 per cent retinoic acid.[8, 9] Retinoic acid is the ligand for about 300 to 1000 genes responsible for growth,[10] differentiation, and maturation of cells [11] (The exact number of genes is not yet known.)

This activity is no surprise because for decades we have known that vitamin A is vital for healthy skin. Wise and Sulzberger[12] suggested in 1938 that the reason that sun-exposed skin is wrinkled whereas sun-protected skin is not, is that the vitamin A in the sun-exposed skin had been damaged by sunlight, and that led to a chronic local hypovitaminosis A in wrinkled skin. At that stage they could not prove it, but later research showed that skin exposed to the sun beyond a certain degree becomes deficient in vitamin A.[13] Furthermore, we learned that even though vitamin A is destroyed by both UVA and UVB,[14] UVA rays are ubiquitous and can penetrate through clouds and windowpanes, so it is easy to understand that RP is easily destroyed every day even in cloudy conditions.

Vitamin A is the one key molecule essential for normal function of all cells of the skin: keratinocytes, melanocytes, Langerhans cells, and fibroblasts.[15] For reasons as yet unexplained, vitamin A has an essential part in controlling pigment deposition.[16] Because vitamin A increases the activity of telomerase in normal cells, it is essential to avoid cell senescence (the ceasing of cell division due to the loss of the ends of chromosomal strands or so-called telomeres).[17] Conversely, vitamin A down-regulates and reduces telomerase enzymes in cancer cells, promoting senescence in these cells.[18]

Professor Cluver was dean of the faculty of medicine in Witwatersrand University when I started studying medicine. Cluver was a pioneer in recognizing that Vitamin A played an essential role in counteracting sun damage,[19] but he only used vitamin A periodically to combat sun damage. He showed that every time we go out into sunlight, the photosensitive vitamin A molecule is denatured not merely in the skin, but also in the blood.[20] The value of RP (retinyl palmitate) was first shown in treating aged skin as long ago as the 1950s![21]

The effects of UV happen very quickly and significantly lower the levels of RP in the skin.[22] Over time, investigations have demonstrated that vitamin A is not only good for ageing skin, but also actually essential.[23] We know that RP specifically protects the DNA from damage and, when applied topically in adequate doses, can have DNA-protective effects to the same extent as a sunscreen of SPF 20.[24] Retinol, retinaldehyde, and retinoic acid do not give this photoprotective effect and actually make skin photosensitive. This is one compelling reason among others for using RP in skin creams rather than the other forms.

Women have an added disadvantage because blood levels of vitamin A drop when they menstruate.[25] That means that they are more vulnerable to photodamage during this part of their menstrual cycle.

The photo-decomposition (destruction by exposure to light) of RP is the main cause of photoageing. This degrading happens day after day and year in, year out.[26] The irony is that paradoxically vitamin A is required to restore the nuclear and cellular receptors, destroyed by the light.[13]

At the same time that vitamin A is being destroyed, antioxidants like vitamin C are also being eliminated by the same process. This daily, localized, deficiency of vitamins A and C and other skin antioxidants is insidious and pernicious. About 10 to 20 minutes of full sunlight will destroy as much as 90 per cent of the cutaneous retinyl esters and it takes longer than 24 hours to restore normal levels.[27, 28]

Retinoid receptors (vitamin A receptors) on the cell membranes are destroyed at the same time and retinoid metabolic pathways in cells become less efficient.[29] Keratinocytes produce less of the essential keratins and ceramides (proteins and fats) that ensure an effective waterproofing barrier for the skin.

***The horny layer becomes much thicker and rougher with a basket weave pattern instead of being compact and thinner, but denser. UV irradiation stimulates the release of matrix-metallo-proteinases (MMP), whereas vitamin A normally inhibits the formation of MMPs secreted by keratinocytes and fibroblasts.[30] With a deficiency of vitamin A, MMPs are released in greater quantities and destroy collagen and anchoring fibrils.

[26, 31] Dermal papillae become flattened.[32] At the same time free radicals are being generated, but unfortunately the important free-radical scavengers of the skin are simultaneously depleted.[33, 34] With acute UV exposure the GAGs increase whereas with chronic UV damage the fibroblast produces fewer glycosaminoglycans so the skin feels drier and wrinkles show up very easily.[35]

Langerhans cells desperately need vitamin A to function normally, and its deficiency prevents them from recognizing DNA damage.[36] As a result, clones of abnormal cells slowly start to develop and, years later, manifest as keratoses (rough sun spots) or skin cancer. Research has shown that low retinyl esters are associated with skin cancers,[37] while adequate use of topical retinoids is a significant preventative measure against skin cancer.[38]

*Before using vitamin A the skin shows the classical signs of vitamin A deficiency: disorganised thick stratum corneum, a poor stratum granulsoum and a thin stratum spinosum with irregular nuclear shapes and staining that affects the basal layer as well. After (5 months) vitamin A the skin is thicker and the layers well defined and the nuclei of the cells are more regular and the stratum corneum is thinner and more compact.*

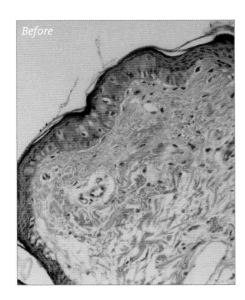
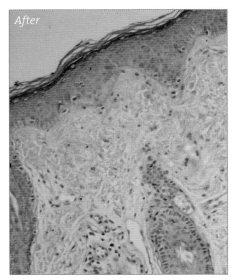

UV and near-UV light stimulates production of melanin, and if there is adequate vitamin A the distribution of melanin in the skin is kept even. Vitamin A (particularly RP) deficiency permits lentigines and melasma to form. Keratinocytes in the stratum spinosum that are enriched with RP will protect the cells below and reduce the UV damage and thereby reduce the chances for melasma and other light-induced skin problems.

Cluver showed that bad sunburn could be improved by oral supplementation of 25,000 IU of RP per day.[39] Conventional sunscreens cannot give adequate protection from UVA, and so vitamin A is still damaged by light, even when a person is wearing a sun protection factor of 30 or 40. On the other hand, application of a vitamin A cream can restore the normal levels within hours. Medical research concentrates on retinoic acid, yet all forms of vitamin A at physiological doses are eventually converted into retinoic acid and are thus equally effective.[40] Naturally though, one has to bear in mind the relative concentrations of the different molecules, as higher concentrations can provide larger numbers of molecules.[41]

In the medical literature, Kligman said that retinyl esters have no clinical value, but unfortunately he made this comment after having looked only at conventional creams containing low-dose retinyl palmitate.[42] Precisely because of the central role of vitamin A and D in normal physiology and because both too little and too much UV light have such dire complications besides melasma and wrinkles, it has become absolutely necessary for us to search for a way to get enough summer sun to make vitamin D and still prevent the problems of photodamage.

Diet plays only a small part in Vitamin D nourishment because it is mainly supplied to us through our skin. However in Europe, the USA, and Japan one will typically get UVB rays in significant quantities for manufacturing vitamin D for only an average of six months of the year. For this reason many people have to supplement their vitamin D as well as spending 10 to 20 minutes in the midday sun every day in late spring, summer, and early autumn. The darker the skin, the more difficult it is to make vitamin D.[43] Unfortunately, the need to stay in the sun for longer poses the risk of sustaining photodamage!

I do believe however, that RP-enriched topical creams with antioxidants are the most effective way, at this stage, to save both the skin and the body at the same time. We have to use the special double action of RP, which protects us from UV light and also ensures that the normal vitamin A physiology is maintained. Of course, most people do not tan every day, but we do still expose some areas to the sun every day, and in those areas we certainly destroy a significant part of the vitamin A that has to be replaced from the stores in the liver.

Let's assume for the sake of argument that 95 per cent of the lost vitamin A is restored and the deficiency is only five per cent (in fact it is greater

> *"Diet plays only a small part in Vitamin D nourishment because it is mainly supplied to us through our skin"*

than that). That is not much, you may say, but let's look at it from a different angle. Imagine that this five per cent is represented by ten dollars. If you lose ten dollars you don't worry. But think of losing ten dollars every day and you will also probably not worry. Let's do the arithmetic and see how bad that really is. At the end of one year, you will have lost $3,650. Not a train wreck, and it is also not so bad at the end of ten years: $36,500! But we live longer than ten years in daylight. We are still children at ten or 11, but we've been exposed to daylight for about ten years. At 20 years we have lost $73,000 and now it looks a little more significant. At 40 when we still want to look young, we have lost $146,000. And the loss continues, and by 80 years of age we have lost $292,000. That is only from losing an insignificant amount every day! Imagine if the amount were more than that!

Look at people who have taken care to restrict their damage by the sun, and compare them with people who have allowed themselves to become sunburned as much as possible. The sunburned people are always older-looking with wrinkles and pigmented blemishes. These people may have enjoyed themselves a great deal but they also built up a significant deficit of vitamin A, and as a result, look older than necessary with large, brown freckles on their skin. If you told them that they had a vitamin deficiency they would laugh at you. In Japan they would point out that their diets are very good. In fact, most Japanese do have a good diet, rich in vitamin A and also carotenoids (plant-derived pre-cursors of vitamin A). Interestingly, oriental skin is richer in beta carotene than white and black skins are. However, the liver can be rich in vitamin A while the skin suffers a chronic vitamin A deficit. This does not make sense because we like to think that our bodies can automatically replace the loss of vitamin in the skin every day, but it simply does not happen that way.

Vitamin A deficiency is probably the most common vitamin deficiency in the world. We get vitamin A from various foods like animal and fish liver oils, and probably the best of all from plants in the forms of the carotenes like beta-carotene, a precursor of vitamin A. The beta-carotene molecule consists of two molecules of vitamin A that can be split by the cells. The form of vitamin A produced is retinaldehyde, which is essential for normal vision. Vegetarians may not suffer from vitamin A deficiency because the body makes as much retinaldehyde from beta-carotene as is required, but they can never have ideal levels of vitamin A such as have been shown to reduce the chances for cancer. Retinaldehyde can be converted into other forms of vitamin A like retinyl palmitate (the storage form), retinol (the alcohol transport form), and retinoic acid (the form that acts on DNA). However, there is a limit, and one cannot take high doses of vitamin A to get the effects of higher vitamin A required, for example, to protect chromosomes from shortening too early.[44]

One important difference between vitamin A and beta-carotene is that a vitamin A molecule can absorb one free radical and then is deactivated and no longer has vitamin A activity. Beta-carotene, on the other hand, can absorb many free radicals and maintain its vitamin A activity.[45] It has been said that beta-carotene can inactivate up to 1000 free radicals because its complex arrangement of double bonds permits it to remain an active free-radical scavenger.

Because it is so essential for vision, you can get night blindness from a deficiency of vitamin A in the retina (the lining of the inside of the eye). Deficiency of vitamin A causes a dry conjunctiva (the outside lining in front of the eyes) and eventually leads to blindness. This condition is called 'xerophthalmia', which is just a Greek word meaning dry eyes.

The most common cause of blindness in the world is vitamin A deficiency, which is easily and completely preventable. Vitamin A deficiency can also cause respiratory infections, loss of the enamel on teeth, reduced senses of taste and smell, weak bones, and chronic diarrhoea. However, these are the signs of well-established, more severe, deficiency. Minor degrees of deficiency may go unrecognized. Little plugs of keratin around the hair follicles of the anterior thighs and posterior arms may go unrecognized in people who do not show the other classical signs of vitamin A deficiency.

You can see that vitamin A is not only important for skin, but also for many other organs of the body. One important system is the immune system. This is why children suffering from measles in third world countries are more likely to die than well nourished people in the first world. By simply giving vitamin A to children we can drastically reduce their likelihood of dying from complications of measles.[46] Now think of this: if generalized vitamin A deficiency is so common and also unrecognized, then what about a local deficiency in our skin? Could we have a chronic, undetected vitamin A deficiency in the sun-exposed parts of our skin?

Vitamin A is necessary for the normal function and structure of our skin.[47] When skin is deprived of vitamin A the horny layer gets thicker and rougher. As mentioned above, the common areas where the thicker horny layer become manifest are the areas around the hair follicles, especially of the back of the arms and the front of the thighs. Paradoxically, more sebum may be secreted, and that in combination with the rough horny layer may cause obstruction of the follicles and the creation of large blackheads. Acne might also result.

With vitamin A deficiency, the growing layer of the skin becomes less active because the cells divide less frequently. The DNA is also affected and damage to the DNA is no longer repaired as effectively as in normal skin. Pigment distribution becomes irregular and darker marks are common. In the dermis the collagen is no longer as healthy as before and the skin loses elasticity from

> *"Vitamin A deficiency is probably the most common vitamin deficiency in the world."*

fewer and poorer quality elastin fibres. There are fewer blood vessels than in normal skin and the skin colour becomes more sallow.

These are exactly the same changes that we associate with ageing, so the idea that vitamin A deficiency is the cause of ageing is not so far fetched after all![48] The best way to make a comparison is to see what happens when we treat aged skin with vitamin A by applying it topically. We see that the horny layer becomes smoother, and while it may get thinner in the first few weeks, it gets thicker with time but remains smoother. Less sebum is made and blackheads become less frequent.[49] Acne, while it may be aggravated initially, diminishes and the skin may even change to look normal. The growing layer of the skin becomes more active, DNA is repaired very efficiently, damaged cells (even cancerous ones) are removed from the skin, and the epidermis gets much thicker. Pigment distribution becomes more even;[16] fewer pigment granules are generally produced. In the dermis, more collagen is laid down and the skin becomes thicker and more elastic. More blood vessels are formed, so the colour of the skin is improved. In fact, the skin takes on all the features of younger skin!

Many people will say that the best way to treat the vitamin deficiency in the skin is to take the vitamin by mouth, but experiments have shown that the vitamin is preferentially stored in the liver and then shared with the heart, the kidney, the bone marrow, the lungs and other tissues. Unfortunately, the skin appears to get the last chance at being fed with the vitamin A. This may help explain why taking large doses of oral vitamin A cannot adequately restore the epidermal and dermal layers of the skin. That is why I believe the best way to replace lost vitamins from the skin is to put them directly on the skin and ensure that they penetrate properly.

## REFERENCES

1. Wolff, A.E., A.N. Jones, and K.E. Hansen, *Vitamin D and musculoskeletal health*. Mol Nutr Food Res, 2008. **4**(11): p. 580-8.
2. Nakamura, K., *Vitamin D insufficiency in Japanese populations: from the viewpoint of the prevention of osteoporosis*. J Bone Miner Metab, 2006. **24**(1): p. 1-6.
3. Gupta, R., et al., *Photoprotection by 1,25 dihydroxyvitamin D3 is associated with an increase in p53 and a decrease in nitric oxide products*. J Invest Dermatol, 2007. **127**(3): p. 707-15.
4. Schwartz, G.G. and H.G. Skinner, *Vitamin D status and cancer: new insights*. Curr Opin Clin Nutr Metab Care, 2007. **10**(1): p. 6-11.
5. Ingraham, B.A., B. Bragdon, and A. Nohe, *Molecular basis of the potential of vitamin D to prevent cancer*. Curr Med Res Opin, 2008. **24**(1): p. 139-49.
6. Kang, S., X.Y. Li, and J.J. Voorhees, *Pharmacology and molecular action of retinoids and vitamin D in skin*. J Investig Dermatol Symp Proc, 1996. **1**(1): p. 15-21.
7. Kurlandsky, S.B., et al., *Auto-regulation of retinoic acid biosynthesis through regulation of retinol esterification in human keratinocytes*. J Biol Chem, 1996. **271**(26): p. 15346-52.
8. KE, H.J.C., *Comparison of the metabolism of retinol delivered to human keratinocytes either bound to serum retinol-binding protein or added directly to the culture medium*. Exp Cell Res, 1998 **238**(1): p. 257-64.
9. Vahlquist, A., *Vitamin A in human skin: I. detection and identification of retinoids in normal epidermis*. J Invest Dermatol, 1982. **79**(2): p. 89-93.

10. Marcelo, C.L. and K.C. Madison, *Regulation of the expression of epidermal keratinocyte proliferation and differentiation by vitamin A analogs.* Arch Dermatol Res, 1984. **276**(6): p. 381-9.
11. Goffin, V., et al., *Topical retinol and the stratum corneum response to an environmental threat.* Skin Pharmacol, 1997. **10**(2): p. 85-9.
12. Sulzberger, M.B. and F. Wise, *The year book of dermatology and syphilology.* 1938 ed1938, Chicago: Year Book Publishers
13. Wang, Z., et al., *Ultraviolet irradiation of human skin causes functional vitamin A deficiency, preventable by all-trans retinoic acid pre-treatment.* Nat Med, 1999. **5**(4): p. 418-22.
14. Berne, B., M. Nilsson, and A. Vahlquist, *UV irradiation and cutaneous vitamin A: an experimental study in rabbit and human skin.* J Invest Dermatol, 1984. **83**(6): p. 401-4.
15. Fu, P.P., et al., *Physiological role of retinyl palmitate in the skin.* Vitam Horm, 2007. **75**: p. 223-56.
16. Ortonne, J.P., *Retinoid therapy of pigmentary disorders.* Dermatol Ther, 2006. **19**(5): p. 280-8.
17. Kosmadaki, M.G. and B.A. Gilchrest, *The role of telomeres in skin ageing/photoageing.* Micron, 2004. **35**(3): p. 155-9.
18. Xiao, X., et al., *Retinoic acid-induced downmodulation of telomerase activity in human cancer cells.* Exp Mol Pathol, 2005. **79**(2): p. 108-17.
19. Cluver, E.H., *Sun-trauma prevention.* S Afr Med J, 1964. **38**: p. 801-3.
20. Politzer, W.M. and E.H. Cluver, *Serum vitamin-A concentration in healthy white and Bantu adults living under normal conditions on the Witwatersrand.* S Afr Med J, 1967. **41**(40): p. 1012-5.
21. Reiss, F. and R.M. Campbell, *The effect of topical application of vitamin A with special reference to the senile skin.* Dermatologica, 1954. **108**(2): p. 121-8.
22. Tang, G., et al., *Epidermis and serum protect retinol but not retinyl esters from sunlight-induced photodegradation.* Photodermatol Photoimmunol Photomed, 1994. **10**(1): p. 1-7.
23. Varani, J., et al., *Molecular mechanisms of intrinsic skin ageing and retinoid-induced repair and reversal.* J Investig Dermatol Symp Proc, 1998. **3**(1): p. 57-60.
24. Antille, C., et al., *Vitamin A exerts a photoprotective action in skin by absorbing ultraviolet B radiation.* J Invest Dermatol, 2003. **121**(5): p. 1163-7.
25. Lithgow, D.M. and W.M. Politzer, *Vitamin A in the treatment of menorrhagia.* S Afr Med J, 1977. **51**(7): p. 191-3.
26. Fisher, G.J., et al., *Molecular basis of sun-induced premature skin ageing and retinoid antagonism.* Nature, 1996. **379**(6563): p. 335-9.
27. Tran, C., et al., *Topical delivery of retinoids counteracts the UVB-induced epidermal vitamin A depletion in hairless mouse.* Photochemistry and Photobiology, 2001. **73**(4): p. 425-31.
28. Andersson, E., et al., *Ultraviolet irradiation depletes cellular retinol and alters the metabolism of retinoic acid in cultured human keratinocytes and melanocytes.* Melanoma Res, 1999. **9**(4): p. 339-46.
29. Andersson, E., et al., *Differential effects of UV irradiation on nuclear retinoid receptor levels in cultured keratinocytes and melanocytes.* Exp Dermatol, 2003. **12**(5): p. 563-71.
30. Fisher, G.J., et al., *Pathophysiology of premature skin ageing induced by ultraviolet light.* N Engl J Med, 1997. **337**(20): p. 1419-28.
31. Varani, J., et al., *Vitamin A antagonizes decreased cell growth and elevated collagen-degrading matrix metalloproteinases and stimulates collagen accumulation in naturally aged human skin.* J Invest Dermatol, 2000. **114**(3): p. 480-6.
32. Li, L., et al., *Age-related changes in skin topography and microcirculation.* Arch Dermatol Res, 2006. **297**(9): p. 412-6.
33. Thiele, J.J., et al., *The antioxidant network of the stratum corneum.* Curr Probl Dermatol, 2001. **29**: p. 26-42.
34. Packer, L. and G. Valacchi, *Antioxidants and the response of skin to oxidative stress: vitamin E as a key indicator.* Skin Pharmacol Appl Skin Physiol, 2002. **15**(5): p. 282-90.
35. Langton, A.K., et al., *A new wrinkle on old skin: the role of elastic fibres in skin ageing.* International Journal of Cosmetic Science, 2010.
36. Murphy, G.F., S. Katz, and A.M. Kligman, *Topical tretinoin replenishes CD1a-positive epidermal Langerhans cells in chronically photodamaged human skin.* J Cutan Pathol, 1998. **25**(1): p. 30-4.

37. Guo, X. and L.J. Gudas, *Metabolism of all-trans-retinol in normal human cell strains and squamous cell carcinoma (SCC) lines from the oral cavity and skin: reduced esterification of retinol in SCC lines.* Cancer Res, 1998. **58**(1): p. 166-76.
38. Lippman, S.M. and R. Lotan, *Advances in the development of retinoids as chemopreventive agents.* J Nutr, 2000. **130**(2S Suppl): p. 479S-482S.
39. Cluver, E.H. and W.M. Politzer, *The pathology of sun trauma.* S Afr Med J, 1965. **39**(41): p. 1051-3.
40. Kang, S., et al., *Application of retinol to human skin in vivo induces epidermal hyperplasia and cellular retinoid binding proteins characteristic of retinoic acid but without measurable retinoic acid levels or irritation.* J Invest Dermatol, 1995. **105**(4): p. 549-56.
41. Duell, E.A., S. Kang, and J.J. Voorhees, *Unoccluded retinol penetrates human skin in vivo more effectively than unoccluded retinyl palmitate or retinoic acid.* J Invest Dermatol, 1997. **109**(3): p. 301-5.
42. Kligman, A.M., *The growing importance of topical retinoids in clinical dermatology: a retrospective and prospective analysis.* J Am Acad Dermatol, 1998. **39**(2 Pt 3): p. S2-7.
43. Yuen, A.W. and N.G. Jablonski, *Vitamin D: in the evolution of human skin colour.* Medical hypotheses, 2010. **74**(1): p. 39-44.
44. Yokoo, S., et al., *Slow-down of age-dependent telomere shortening is executed in human skin keratinocytes by hormesis-like-effects of trace hydrogen peroxide or by anti-oxidative effects of pro-vitamin C in common concurrently with reduction of intracellular oxidative stress.* J Cell Biochem, 2004. **93**(3): p. 588-97.
45. Sies, H. and W. Stahl, *Vitamins E and C, beta-carotene, and other carotenoids as antioxidants.* Am J Clin Nutr, 1995. **62**(6 Suppl): p. 1315S-1321S.
46. Sommer, A., *Xerophthalmia and vitamin A status.* Prog Retin Eye Res, 1998. **17**(1): p. 9-31.
47. Fisher, G.J. and J.J. Voorhees, *Molecular mechanisms of retinoid actions in skin.* FASEB J, 1996. **10**(9): p. 1002-13.
48. Saurat, J.H., *Skin, sun, and vitamin A: from ageing to cancer.* J Dermatol, 2001. **28**(11): p. 595-8.
49. Griffiths, C.E., *Drug treatment of photoaged skin.* Drugs Ageing, 1999. **14**(4): p. 289-301.

# Chapter 09
# MY PERSONAL EXPERIENCE WITH USING VITAMIN A-BASED SKIN-CARE PRODUCTS.
# DR DES FERNANDES

I used vitamin A in its medical form on my own skin in 1982 by mistake. By then I had been using vitamin A on patients as retinoic acid from as early as 1977 to attempt to improve scars and also to treat acne. I had noticed that my patients developed surprisingly beautiful skin. So in 1982 when I heard that my friend, the wife of my plastic surgery teacher, was using retinoic acid on her skin to make herself look younger, I decided I wanted to do the same thing.

However, my skin turned out to be very sensitive to vitamin A in the retinoic acid form. It is possible that I have a deficient system for manufacturing retinoid receptors, for I developed a retinoid reaction that lasted for several months. A few years later when I met my friend again, I had changed to retinyl palmitate creams and she remarked on how good I was looking. I replied that it was thanks to her, but ironically it turned out that she had never actually used retinoic acid and did not know what I was talking about.

I realized the significance of alternative forms of vitamin A by 1986 as a result of extensive reading about the metabolism of vitamin A when I was researching malignant melanoma. My reading led me to the surprising discovery that topical vitamin A had been used to treat solar keratoses and even small basal cell carcinomas as early as 1960. Subsequent articles showed the value of topical vitamin A in normalizing sun-damaged skin. I realized then that topical vitamin A was going to become more and more important because the climatic changes the world was experiencing. Because not everyone would be able to get the medical form of vitamin A, I decided that the cosmetic forms should be promoted. I believed that through the natural conversion of the forms of vitamin A by skin cells, the cosmetic form would be changed into the medical form inside the cell. There was already a wealth of research information available on vitamin A in medical research literature. The bulk of the data pointed to the fact that topical vitamin A was the best way to treat skin excessively exposed to ultraviolet light.

On further research I discovered information about beta-carotene, vitamin C, and vitamin E, which were each shown to be of value in reducing free-radical damage in sun-exposed skin. I thought that this information was so important that someone would have already made a cosmetic cream containing adequate doses of vitamin A, C, E, and beta-carotene. Since I wanted this cosmetic for my patients, I hunted for a cosmetic that could match and satisfy these requirements, but to my disappointment I found nothing anywhere that even came close. I then approached a cosmetic company and suggested that they should start to make such cosmetics with these particular pharmaceutical properties and functions. However, they ignored my request and I realized that I would simply have to make these products myself if I wanted them.

The other important aspect I considered at the time was that these cosmetics would have to be attractive to young people if we were going to prevent the ravages of sunlight in later life. For real prevention later on we would have to start early. I commenced making products in my own kitchen, but I certainly did not make any really good creams.

Later I worked with a trained chemist, and that was the start of a range of serious cosmetic products, which were then made in Cape Town, South Africa. Here are some photographs of patients

*GM before and after using vitamin A*

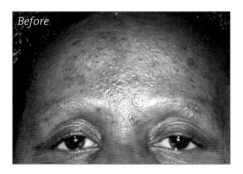
*Before*

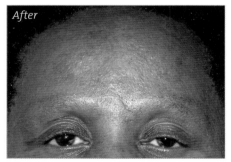
*After*

that I treated with the early, pioneering creams that I formulated and made of vitamin A, beta-carotene, and vitamins C and E and pro-vitamin B5.

I wanted to be as gentle on the skin as I could, and so we made creams without any added preservatives or perfumes. I also used low-dose alpha-hydroxy acids with the aim of enhancing the penetration of the vitamins. The only preservatives would be those trace amounts included in the raw materials to preserve them in the best possible state before manufacturing. This meant that the creams had a mild, not-unpleasant, smell from the ingredients. Of course, such fresh product also meant that an open tube was effective for about three months only.

Although the formulations have developed greatly in the past 20 years and have become much more sophisticated and evolved, the basic principles laid down right at the beginning are still being maintained in every formulation.

I will show examples of various skin problems that we deal with by age. Young people have to learn that they should start using vitamin A and antioxidants as soon as they can. Several people who have listened to my philosophy put their children on Environ vitamin A creams from about the age of two years. The interesting fact is that these children have climbed up the 'vitamin A ladder' (gradual increase of the vitamin A dosage to the skin) easily and were able to use even the highest level of vitamin A without developing a retinoid reaction. I believe the reason for this is that at a young age, with limited sun exposure, they still have many retinoid receptors on their cell walls, so they absorb the vitamin A into the cells rapidly. They probably also produce larger populations of receptors in response to higher doses of vitamin A more easily.

As the years have passed I have thought about our experience with the topical vitamins and the advice that we have given people. I used to believe that as we got older we should use higher levels of vitamin A, and young people did not need such high levels. This is in fact incorrect, because it presupposes that younger people sustain less

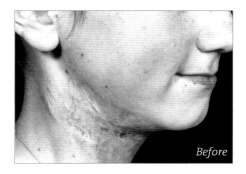
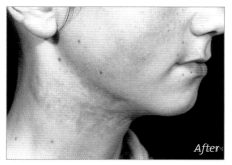

*10 year old hot water burn scars on neck treated with medium dose vitamin A once a day and an SPF 4 sunscreen. Result after 5 months shows changes that we can interpret as TGF-beta-3 activity.*

> "I believe that by keeping the vitamin A levels high in the skin we will reduce the chances of these children developing severe acne."

damage to the vitamin A in their skin than older people, but younger children are, generally speaking, more exposed to the sun than older people. They too need the vitamin A as retinyl palmitate, as a natural sunscreen, and to maintain the normal activities of vitamin A. Their need is at least as great as that of an older person. Furthermore, I believe that by keeping the vitamin A levels high in the skin we will reduce the chances of these children developing severe acne. In later life continued use of vitamin A should lead to less patchy, less pigmented, and less saggy skin, because the skin volume of collagen will be much greater and the stimulus to produce melanin will be less.

This is a photograph of a 12-year-old girl. Most people would believe that she is too young to use a cosmetic, but it is important to remember that if you want to preserve young skin, you have to start when the skin is young. That way you can have a beautiful skin for a lifetime. She was accidentally scalded with hot water when she was two years old. (Scalding is a common injury in children.)

The scar depicted in the first picture is already ten years old and has improved as much as it can on its own. If you study the photographs carefully, you will see that she already shows signs of photoageing, even at ten.

Look at the elastosis on her neck. This is not as surprising as it may seem, because when we critically analyse sun damage throughout our lives, we find that about 80 per cent of damaging sun exposure happens before we reach the age of 20. Note that she also has whiteheads on her cheek, these being the earliest signs of the obstruction of sebaceous follicles, and they will eventually lead to blackheads or acne.

Of great concern to her was that her brothers had started with whiteheads on their skin in the same way, and they had developed into bad acne. The skin scars evident in the picture show twisted whorls of scar collagen, an irregular surface, and pigment blemishes. I recommended that she use a very simple cleaning regime with a very mild cream cleanser and then tone up the skin with an alpha-hydroxy acid toner. She was then recommended to use a simple day cream of vitamin A in a mild dose, antioxidants and a sun protection cream SPF 4. As she really liked to play tennis, I suggested that she should apply a stronger antioxidant-enriched sunscreen of SPF 16 before playing. At that time, the sunscreen we made was the only antioxidant-enriched sunscreen in the world; the girl was in fact using products that she could get nowhere else.

Her treatment consisted of this very simple regime, and she used nothing else on her skin. She came back five months after the start of treatment for some photographs. Incidentally, we were still using high-definition film cameras at that time, as the digital technology available then could not provide the detail I required. The photographs show typical changes that I have learned to expect from efficient use of vitamin A and the antioxidants.

① First of all, the normal skin surface is smoother, the whiteheads have disappeared, and the skin looks lovely and fresh.

② The elastosis nodules were small before but now they are almost invisible.

③ The scarred skin has also improved. The scar surface is smoother and the whorls of collagen are less noticeable.

④ Most importantly though, the pigment blemish has almost disappeared as the result of the action of vitamin A on the melanocytes and also the important action of the antioxidants. Melanin formation requires oxidation, and if the melanocytes are kept rich in antioxidants, then melanin formation is impeded. Vitamin C has a very important role to play here. The pigment formation was further improved as a result of using the simple day cream SPF 4 and using the stronger sunscreen when playing tennis.

I honestly think that I changed this young girl's life and made it better. It is impressive that these results were achieved by her own body working with the correct, safe, effective, vitamins in sufficient doses. It is also interesting that even though she applied the products only during the day, she still had a very good and valuable result.

## LATE TEENAGERS: ACNE

The big problem for people in their late teens is the development of acne. This girl shows the oily skin with large pores, blocked follicles, blackheads, and acne typical of this problem. She should have lovely skin, and that is what she got after using a very simple regime of cleaning her skin gently with a gel cleanser and toning with a stronger alpha-hydroxy acid toner. She then used a gel formulation of vitamin A, beta-carotene, and the antioxidants only once per day. The patient reported that the change took about two months, while the photograph shows the result five months later.

To date I have found that this regime works on more than 70 per cent of cases. Other cases that are not responding as well require either better cleansing of the skin prior to washing by using mineral-oil-derived oil cleansers and

> "I honestly think that I changed this young girl's life and made it better."

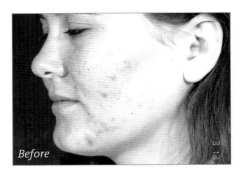
Before

After

*After five months continuous use of vitamin A in moderate doses, antioxidants and mild alpha hydroxyacid creams the skin is clear and healthy looking.*

*Acne can be extremely difficult and in this case we persisted with vitamin A, tea tree oil, and salicylic acid and ultimately after two years she developed healthy looking skin. (my pictures show a clear skin whereas it seems blotchy here).*

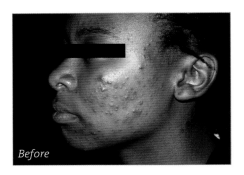
Before

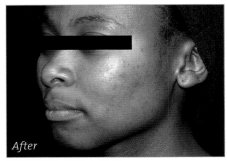
After

make-up removers, more alpha-hydroxy or beta-hydroxy acids, or antiseptics like Australian tea tree oil and other ingredients that reduce sebum production.

If that fails, then one needs to consider the advantages of skin peels. Diet generally does not play an important part, but by increasing the daily oral intake of vitamin A by 5000 to 10,000 IU or even 40,000 IU per day, acne may also improve. Failing that, special forms of oral vitamin A should be used like cis-retinoic acid. No one today should have uncontrolled acne and get permanent scars and pits on the face.

In recent years my approach has been refined to using the light peels right from the beginning. I teach patients to have one or two peels with a therapist and then very short ten-minute daily home peels followed by the application of the vitamin A, C, and E oil.

### DISTURBED SKIN

When I first heard that a friend of mine, Dr Akiko Tozawa in Tokyo, was using vitamin A and the beta-carotene antioxidant cream on people with disturbed skin like atopy (allergic eczema), I said that she should not confuse a medicine with a cosmetic. However, I had forgotten about the fact that vitamin A is the 'skin normalizer'. I should not have been surprised that it can be used in many people to restore health to their skin, especially when such skin has been damaged by cortisone.

I also treated people with allergic skin conditions, and I was surprised by the changes that I saw. I advised patients that they could stop using cortisone. Here are some photographs of people who suffered from skin atopy and were treated with vitamin A, beta-carotene, and the antioxidants vitamins C and E and pro-vitamin B5. The skin sometimes is so bad that it is difficult to believe it will ever be normal again, yet with patient application of the advance vitamin topical treatment, the repair happens.

### MELASMA

Pigmentation of the skin is the most challenging problem. No one in the world has the complete answer and

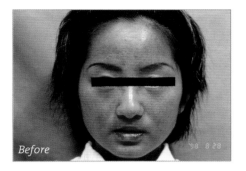
*Before*

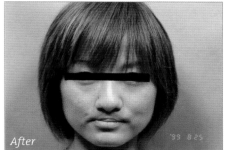
*After*

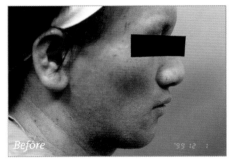
*Before*

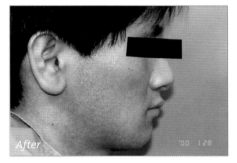
*After*

*Atopy presents with an unstable stratum corneum and since vitamin A has a dominant role in the growth of keratinocytes to produce a healthier stratum corneum, it also has an important role in treating patients with atopy. These two patients demonstrate this effect and the improvement continues as long as vitamin A is used topically every single day.*

*Atopy in a young Japanese man is settled by stopping topical cortisone and substituting with vitamin A and antioxidant gel twice a day. It is recommended to stay permanently on this regime even though the skin clears remarkably.*

it continues to plague our clients and defy most treatments. However, I find some surprising results after using just vitamin A and antioxidants.

This woman joined a trial on skin care and she never came back. I thought she did not like the cream I gave her. However, she really liked it but she did not want to come to my clinic because of the heavy traffic, so she decided to buy the cream instead. When I saw her again the pigmentation on her cheek was almost invisible. Over the subsequent years her skin has improved even further. I found similar results from treatment of chloasma.

Lentigo (sun freckles) is difficult to treat because we are always exposing our skin to sunlight, and it seems that melanin production in affected skin has become hypersensitive to light. Minimal amounts of light can now trigger off the increased pigment system. Pigmentation formation is mostly initiated by exposure to sunlight, the production of free radicals, and activating factors from the keratinocytes, which stimulate the melanocytes to produce more melanin.

Another problem in treating lentigo lesions is that part of the pigmentation is not due to melanin but is from lipofuscin. This is a yellow-brown pigment left over from the breakdown and absorption of damaged blood cells. As one ages one produces more lipofuscin; called the ageing pigment, it commonly causes 'age spots' on the hands and face.

*Many people with sensitive skin believe that vitamin A will irritate their skin. Here is a good example of a client gradually building up from low doses of vitamin A to moderately high doses of topical vitamin A, without suffering from a retinoid reaction. The normalization of skin surface and colour is a normal expectation when using topical vitamin A. The skin appears normal, healthier and younger.*

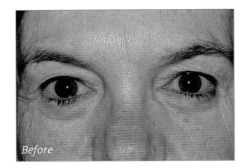
Before

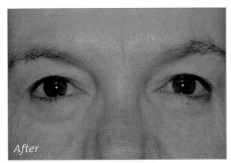
After

My strategy in treating age spots or lentigo is as follows:

(A) Clothing and hats to prevent sun damage;
(B) Reducing production of tyrosinase;
(C) inhibition of tyrosinase;
(D) Competitive block of tyrosinase;
(E) Reduction of melanin transfer into keratinocyte; and
(F) An effective sun-protecting cream predominantly formulated to contain reflectors of UV and other light, and anti-oxidants to control exposure to sunlight.

I have tried a number of products and some have been very successful but are difficult to use, however we are gradually getting closer to improving the skin of many patients.

### PHOTOAGEING

This lady was only two years older than I, but she looked much older when she first came to see me. This is typical chronic severe deficiency of vitamins A and C and the other antioxidants that comes from excessive exposure to sunlight. This is very common in South Africa and used to be considered an inevitable part of maturing. The next stage would be the development of solar keratoses (scaly whitish areas of thickened epidermis) and eventually skin cancer. She had a

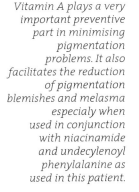

*Vitamin A plays a very important preventive part in minimising pigmentation problems. It also facilitates the reduction of pigmentation blemishes and melasma especialy when used in conjunction with niacinamide and undecylenoyl phenylalanine as used in this patient.*

Before

After

*Before*

*After*

*Melasma in this case was treated with only vitamin A and antioxidants with beta-carotene. Changes seen one year later.*

sallow appearance, rough skin with a thickened horny layer and wrinkles.

I recommended that she clean her skin with a mild cream cleanser, tone with an AHA toner, and then progress from a mild range of vitamin A day and night creams to a medium level of vitamin A creams. Periodically she also used a mild alpha-hydroxy acid cream to help hasten the smoothing of her horny layer.

When I saw her five months later she told me that initially she thought that the creams had worked for a while but that her skin had returned to the same condition as before. I pulled out her original photographs and she burst into tears. She said she had no idea that her skin had been that bad. Now this was actually an interesting change to watch. She had dry skin to start with, and when she started to use vitamin A her dry skin got worse, and she even thought that she should stop the creams. However, in her case, the natural moisturizing factors quickly increased and her skin became normal in hydration. The other changes of vitamin A are easy to see in this case: (a) thicker, healthier skin; (b) better colour; (c) fewer wrinkles.

Her skin became more normal. I showed these same photographs to professors in the USA, and they asked me to provide an affidavit to confirm that she had not had a facelift. Unfortunately,

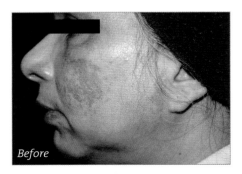
*Before*

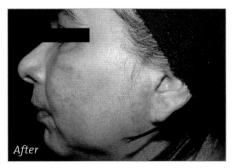
*After*

*Melasma treated with creams twice a day containing vitamin A in stepped up doses, and antioxidnats enriched with extracts of rooibos, honeybush and green tea with resveratrol. Persistent use leads to gradual but continuous reduction of the pigmentation.*

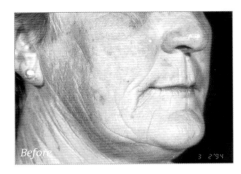
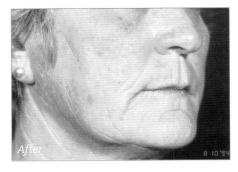

*This patient was one of the very first in the world to photographically prove that topical retinyl palmitate in moderate doses gave results similar to using retinoic acid. We see changes after five months use morning and evening. The patient thought that there had been little change until she saw her pictures and she burst into tears.*

I could not do that because a few weeks before she was due to see me, she was killed in a motor accident.

In the past few years I have found that abnormal pigmentation is improved by medical needling in combination with the application of the topical vitamins. What is probably happening in this situation is a 're-instruction' of the pigment partnership by the reparative cell signals, which were set free by the blood platelets during the needling treatment. This makes sense on a biochemical level.

### PHOTOAGEING OF THE ARMS

I found that we can also get good changes in the arms, but when treating the body I prefer to use a different AHA in combination with vitamin A.

This lady was my piano teacher but she liked to spend time in the sun in the garden and so her arms became badly damaged.

One can see the state of her arms before we started her on vitamin A and a lotion of neutralized lactic acid. Lactic acid helps to restore skin moisture and also smoothes the surface layer of the skin by improving the horny layer. Eight months after starting to use the combination of vitamin A and neutralized lactic acid, her skin has changed noticeably. Of course, it is not wonderful skin for a 74-year-old, but it is a lot better than it was.

*Its not too late to make significant changes. This septuagenarian with marked photodamage of her arms responded to retinyl palmitate oil and ammonium lactate to get significant improvement in eight months.*

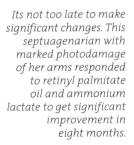
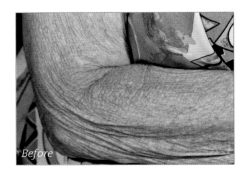
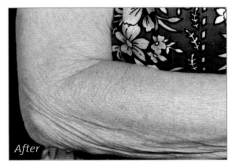

## SUMMARY

In summary I can say that we all need vitamin A in our skin, but we all damage the vitamin A that is originally there. Therefore, science has taught us three important lessons:

1. Vitamin A is the normalizer of skin and skin that remains rich in vitamin A throughout the day will function well and limit the photodamage.
2. We need to replace the vitamin A topically every single day if we want to keep our skin in optimum health.
3. To preserve young skin for as long as possible, we need to start treating young people before their skin gets irreparably damaged. Prevention is, after all, much better than cure.

From my experience I can say that even when it seems that vitamin A does not suit your skin, you actually need it on the skin. You should persist and use very low doses of vitamin A, and eventually your skin will become more normal. I have no doubt about that.

Another way to introduce vitamin A is to take vitamin A supplements to build up your natural level of retinoid receptors so that your skin can once again be nourished by this most essential vitamin A. I believe that people all around the world, of all colours need to apply vitamin A every day to their skin, and that the people who cannot do so are condemned to experience most of the problems of photodamage.

When one truly understands the activities of vitamin A, one questions whether anyone can afford NOT to use topical vitamin A. I have used it myself since 1982 and my skin is constantly admired because it just looks natural and relatively undamaged. I wish I had started at age two!

# Chapter 10
# THE SAFETY OF VITAMIN A BY MOUTH AND APPLIED ON THE SKIN, BY DR DES FERNANDES

Some people read about health and discover that vitamin A can be toxic to the liver, and also cause bad skin. They become alarmed when they read that vitamin A could also cause foetal defects. Then they read a chapter like this and think I am a dangerous man because I am persuading people to put a 'toxic' substance on their skin.

Before we go any further I should explain an important principle in toxicology, the study of poisons. Paracelsus, now regarded as the father of toxicology, was born in 1493, but what he said then rings true today. In German he wrote: *Alle Ding' sind Gift, und nichts ohn' Gift; allein die Dosis macht, daß ein Ding kein Gift ist*. ('All things are poison, and nothing is without poison; only the dose permits something not to be poisonous.') Or, more commonly: 'The dose makes the poison.' That is to say, substances considered toxic are often harmless in small doses, and conversely an ordinarily harmless substance can be deadly if over-consumed. This is a general rule with some exceptions.

Incidentally, everything is a chemical: water is a chemical, oxygen, insulin, oestrogen, and every other substance one can name. Thus when someone tells you that they produce chemical-free products, don't use toxic chemicals and only use natural substances, then you know either they are poorly informed or they are trying to deceive you.

Professor Bruce Ames was professor of biochemistry and molecular biology at the University of California, Berkeley, and a senior scientist at Children's Hospital, Oakland Research Institute. He is the inventor of the Ames test, a system for easily testing the safety of compounds. He has a very sober view on the dangers of chemicals, which is worth examining at this point[1] and quoting at length:

❶ 'The major causes of cancer are as follows:
   (A) smoking: about a third of U.S. cancer (90 per cent of lung cancer)
   (B) dietary imbalances, e.g. lack of dietary fruits and vegetables: the quarter of the population eating the least fruits and vegetables has double the cancer rate for most types of cancer compared to the quarter eating the most; micronutrients may account for much of the protective effect of fruits and vegetables. Excess calories may also contribute to cancer.
   (C) Chronic infections: mostly in developing countries.
   (D) Hormonal factors influenced by life-style.

❷ 'There is no epidemic of cancer, except for lung cancer due to smoking. Cancer mortality rates have declined 16 per cent since 1950 (excluding lung cancer and adjusted for the increased life span of the population).

❸ 'Regulatory policy that is focused on traces of synthetic chemicals is based on misconceptions about animal cancer tests. Recent research contradicts these ideas:
   (A) Rodent carcinogens are not rare. Half of all chemicals tested in standard high-dose animal cancer tests, whether occurring naturally or produced synthetically, are 'carcinogens'.
   (B) There are high-dose effects in these rodent cancer tests that are not relevant to low-dose human exposures and which can explain the high proportion of carcinogens.
   (C) Though 99.9 per cent of the chemicals humans ingest are natural, the focus of regulatory policy is on synthetic chemicals. Over 1000 chemicals have been described in coffee: 27 have been tested and 19 are rodent carcinogens. Plants that we eat contain thousands of natural pesticides, which protect plants from insects and other predators: 64 have been tested and 35 are rodent carcinogens.

❹ 'There is no convincing evidence that synthetic chemical pollutants are important for human cancer. Regulations that try to eliminate minuscule levels of synthetic chemicals are enormously expensive: EPA estimates that total expenditures on environmental regulations cost $140

> "*Incidentally, everything is a chemical: water is a chemical, oxygen, insulin, oestrogen, and every other substance one can name.*"

> billion/year. It has been estimated by others that the United States spends 100 times more to prevent one hypothetical, highly uncertain death from a synthetic chemical than it spends to save a life by medical intervention. Attempting to reduce tiny hypothetical risks also has costs; for example, if reducing synthetic pesticides makes fruits and vegetables more expensive, thereby decreasing consumption, then cancer will be increased.
>
> ❺ Improved health will come from knowledge due to biomedical research and from life-style changes by individuals. Little money is spent on biomedical research or on educating the public about lifestyle hazards, compared to the cost of regulations.

'It is clear from these facts that scare mongering is a dangerous and costly practice, which endangers people's lives rather than improving their quality of life. The Internet has made it easy for people to cast unfounded doubt on very valuable resources, thereby leading to harmful avoidance practices. Doubt sticks very easily in people's minds and they may be given to dangerous or harmful habits as a result. It often takes super-human effort to turn such a trend around. One should be very careful of so-called facts that are presented on the Internet as Gospel truths.'

Let's return now to the all-important matter of the safety of vitamin A. We mentioned Ames just to try and get a perspective on public perception and the reality of vitamin A and cancer. In Arizona, for instance, Alberts found that: 'A placebo-controlled trial in 2297 randomized participants with moderately severe actinic keratoses wherein 25,000 IU/day vitamin A caused a 32 per cent risk reduction in squamous cell skin cancers. We hypothesized that dose escalation of vitamin A to 50,000 or 75,000 IU/day would be both safe and more efficacious in skin cancer chemoprevention. One hundred and twenty-nine participants with severely sun-damaged skin on their lateral forearms were randomized to receive placebo or 25,000, 50,000, or 75,000 IU/day vitamin A for 12 months. The primary study end points were the clinical and laboratory safety of vitamin A, and the secondary end points included quantitative, karyometric image analysis and assessment of retinoid and retinoid receptors in sun-damaged skin. ('Karyometric' relates to the counting or quantitative analysis of cell nuclei in a tissue sample.) Results:

❶ There were no significant differences in expected clinical and laboratory toxicities between the groups of participants randomized to placebo, 25,000 IU/day, 50,000 IU/day, and 75,000 IU/day.

❷ Karyometric features were computed from the basal cell layer of skin biopsies, and a total of 22,600 nuclei from 113 participants were examined, showing statistically significant, dose-response effects for vitamin A at the 25,000 and 50,000 IU/day doses. These karyometric changes correlated with increases in retinoic acid receptor alpha, retinoic acid receptor beta, and retinoid X receptor alpha at the 50,000 IU/day vitamin A dose.

Conclusions: The vitamin A doses of 50,000 and 75,000 IU/day for one year proved safe and also more efficacious than the 25,000 IU/day dose and can be recommended for future skin cancer chemoprevention studies.'[2]

In this study the amount of vitamin A refers to oral supplementation, in other words vitamin A taken in pill or capsule form. It is true that if you eat too much vitamin A you can poison your liver, but one has to eat some very specific and rather unusual foods to be able to achieve this. The Inuit (also called Eskimo) are the people who are most likely to get vitamin A poisoning from eating polar bear liver, which is exceedingly rich in vitamin A. How many people do you know who eat polar bear liver?

However, there are people whose own livers and kidneys do not work properly because of disease, and these people must consult their doctors before taking any vitamin A and D supplements. Another group of people who can get liver toxicity from too much vitamin A are those who consume vitamin tablets excessively and ingest more than about 500,000 IU of vitamin A every day (over 100 times the RDA).[3] Symptoms and effects of excessive vitamin A ingestion are usually reversible on cessation of overdosing.

Factors influencing chronic hypervitaminosis A (overdosage) include the amount taken or dosing regimen; the physical form of the vitamin A; the general health status of the person; other dietary factors such as alcohol and protein intake; and interactions with vitamins C, D, E, and K, the other fat-soluble vitamins.

Deficiency, on the other hand is just as bad and can result in xerophthalmia (dry eyes) and permanent blindness. Too little vitamin A is associated with increased death rates among children. Both excess and deficiency of vitamin A in pregnant animals was shown to cause birth defects (teratogenic). In humans, congenital malformations associated with the pregnant mother's overuse of high doses of vitamin A were reported, but no cause-and-effect relationship has ever been established for certain. Deficiency of the vitamin during pregnancy, on the other hand, has been associated with congenital abnormalities. There is concern about the proper dose of vitamin A during pregnancy. We have to remember that embryos require vitamin A in order to grow, differentiate (specialize their different tissues), and mature. Studies were performed in the cynomolgus monkey (Macaca fascicularis) to provide risk-assessment information on safe dose levels of vitamin A during human

> *"The epidermis does not have a direct blood supply, and is therefore even more likely to suffer vitamin A deficiency."*

pregnancy. This approach yielded safe levels of vitamin A during pregnancy in the range of approximately 25,000 to 37,000 IU/day. [4, 5]

Reported incidences or cases of vitamin A toxicity are rare and have averaged fewer than ten cases per year from 1976 to 1987 in the USA with its very large population. In contrast thousands more people—maybe millions of people worldwide—suffer from night blindness as a result of vitamin A deficiency. This is a case in point where misconceptions and scaremongering not based on any fact are doing immeasurable harm, often to vulnerable and poor, uneducated people.

When I learned enough about vitamin A in the early 1990s, I started to take vitamin A and have done so every day since then. Some anti-ageing doctors take 100,000 IU every day as part of an anti-ageing strategy. Even for people taking vitamin A by mouth, I still recommend the application of topical cosmetics containing vitamin A and antioxidants. The reason for this, as was mentioned earlier, is that the body stores the vitamin A and distributes it to organs in descending order, so that the skin receives only a small amount in the end. The epidermis does not have a direct blood supply, and is therefore even more likely to suffer vitamin A deficiency.

As for the safety of vitamin A taken orally, although most people and most doctors are concerned about 'high doses' of vitamin A – by which they mean 10,000 IU – there are other doctors who believe that 10,000 IU is a low dose for normal people. To get the ideal effects of vitamin A we should take in the region of 40,000 IU daily.

## TOPICAL VITAMIN A

When we apply vitamin A to the skin, only a fraction of it actually penetrates down to the level of the dermis, and the most sophisticated modern tests have barely been able to detect any vitamin A absorbed into the bloodstream when very high doses are applied to the skin. [6] It seems that the skin likes to keep all the vitamin A that it can get and does not easily allow it to be absorbed into the bloodstream. Repeated tests have confirmed that it is very safe to use vitamin A on the skin and there is not the remotest chance of topical vitamin A ever causing any damage to the skin, the liver, or other organs. [7, 8]

The same is true for a pregnancy. Topical vitamin A can never ever give the body the levels of vitamin A that are required to produce a foetal abnormality. The likely situation is that any vitamin A that is absorbed from the skin is only a fraction of the normal amount of vitamin A that is absorbed from the diet. This would in turn be a very small fraction of the amount that a pregnant woman needs as a daily supplement to ensure her health and that of the foetus. Moreover, extra vitamin A in the right amount would only be good for the foetus at that level.

It is therefore safe to use vitamin A creams and gels on the face and body skin when you are pregnant without worrying about the development of the foetus. The *possible* exception to this rule is that of cream or gel containing retinoic acid but there has never been a proven case of any interference with a foetus.

If you are the type of person who wants to make extra, extra sure, then avoid using all alcohol during pregnancy and smoking or even being in the same room where someone is smoking, and beware of a diet deficient in vitamin A and folic acid. If you prefer, you can also avoid using topical vitamin A in the first eight to 12 weeks of your pregnancy. I sometimes wonder why people worry about something as harmless as topical vitamin A and seem to forget totally about the dangers to the baby of tobacco, alcohol, and other pollutants which are proven to be really harmful to unborn human babies.

**Important Note:** Many cosmetics companies do not promote the use of their products during pregnancy in spite of the safety described above, simply because they wish to avoid any unfounded or malicious litigation based on the misconceptions and misinterpretations about vitamin A in pregnancy that are rife in the world. This important reason should not be misinterpreted as an admission of any risk, but is merely a practical measure within the framework of sound business. No one need have any fear of safety for a foetus after inadvertent application of any of the products.

## SUMMARY

In summary it is completely safe to use even high doses of vitamin A on the skin because virtually none of it is absorbed into the bloodstream. Therefore topical vitamin A as retinyl palmitate, retinyl, acetate, retinol, retinaldehyde and even retinoic acid is safe to use. When we consider oral vitamin A, what we have learned is that in healthy people the doses are safe at much higher levels than generally perceived. In time people will use higher and higher doses of vitamin A to get the other benefits of it in keeping us protected from cancers and lengthening the effective lifespan of our cells.

## REFERENCE

1. Ames, B.N. and L.S. Gold, *The causes and prevention of cancer: the role of environment.* Biotherapy, 1998. **11**(2-3): p. 205-20.
2. Alberts, D., et al., *Safety and efficacy of dose-intensive oral vitamin A in subjects with sun-damaged skin.* Clin Cancer Res, 2004. **10**(6): p. 1875-80.
3. Bendich, A. and L. Langseth, *Safety of vitamin A.* Am J Clin Nutr, 1989. **49**(2): p. 358-71.
4. Hendrickx, A.G., et al., *Vitamin A teratogenicity and risk assessment in the macaque retinoid model.* Reprod Toxicol, 2000. **14**(4): p. 311-23.
5. Mastroiacovo, P., et al., *High vitamin A intake in early pregnancy and major malformations: a multicenter prospective controlled study.* Teratology, 1999. **59**(1): p. 7-11.
6. Kochhar, D.M. and M.S. Christian, *Tretinoin: a review of the nonclinical developmental toxicology experience.* J Am Acad Dermatol, 1997. **36**(3 Pt 2): p. S47-59.
7. Van Hoogdalem, E.J., *Transdermal absorption of topical anti-acne agents in man; review of clinical pharmacokinetic data.* J Eur Acad Dermatol Venereol, 1998. **11 Suppl 1:** p. S13-9; discussion S28-9.
8. Krautheim, A. and H. Gollnick, *Transdermal penetration of topical drugs used in the treatment of acne.* Clin Pharmacokinet, 2003. **42**(14): p. 1287-304.

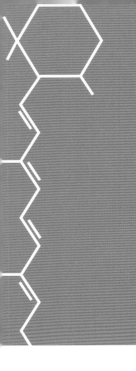

# Chapter 11
# ANTIOXIDANT VITAMINS: C, E, AND OTHERS

**VITAMIN C**

The chemical name for vitamin C is ascorbic acid and it is water-soluble and abundantly found in nature. While some animals have the ability to produce their own vitamin C, human beings do not produce it themselves and are totally reliant on the vitamin C they get in their diets. We do not store large amounts of vitamin C in our bodies, and if we do not ingest enough, signs of vitamin C deficiency, or scurvy, develop within about six weeks.

Ascorbic acid is one of the most important vitamins in our lives. The exact chemical configuration of natural vitamin C is described as L-ascorbic acid, which is the only form found in nature. The 'L-' in front of the name is a chemical description (levorotary or 'turned to the left'), which means that if polarized light is passed through ascorbic acid crystals, the light is twisted to the left. When vitamin C is synthesized equal amounts of D-ascorbic and L-ascorbic acid are created. As you have probably guessed, it is labelled 'D-' because it twists light to the right (dextrorotary or 'turned to the right'). Because synthetic ascorbic acid comprises both D- and L-forms it is designated as D-L-ascorbic acid. Fortunately they can be separated without too much expense.

It just so happens that virtually all 'natural' L-ascorbic acid that is used in the manufacture of cosmetics or oral supplements, is synthesized, but it has been 'purified' and 'made natural' by excluding the D-ascorbic acid. Although this ascorbic acid is synthesized, it is identical to vitamin C found in nature and can thus be called Nature Identical.

Vitamin C is normally deposited in the skin and is an essential part of the antioxidant brigade to protect skin against free radical assault from the atmosphere and from ultraviolet light. Vitamin C plays a very important role in converting deactivated vitamin E back into an active antioxidant form.[1] This is probably the reason that vitamin C plays such an important part in the protection of cellular membranes even though it is water-soluble, while cellular membranes are predominantly composed of lipid molecules. Vitamin C is denatured by heat and UVA exposure. As with vitamin A, we probably develop a chronic deficiency of vitamin C in all the areas of skin that are frequently exposed to sunlight. However, under normal conditions vitamin C is fairly rapidly restored to its normal, physiological level.

Vitamin C seems to have limited activity on DNA itself but stimulates at least four genes related to collagen and in particular elastin production.[2-4] It is absolutely necessary for the production of normal collagen[5] because it plays an essential part in the incorporation of the amino acid proline into collagen and is intimately involved in the formation of elastin.[6] Since proline is essential for normal healthy collagen, a deficiency of vitamin C results in imperfect collagen and the skin tends to become more wrinkled and lose tensile strength, and then it sags. The supplementation of vitamin C by mouth as well as topically boosts the manufacture of new collagen, and so wrinkles become less noticeable.

Another important protective function of vitamin C is to ensure the migration of fibroblasts in the repair of wounds and DNA.[7] When they contracted scurvy the sailors of old suffered non-healing wounds and died easily of infection.

## SPECIFIC ACTIVITIES OF VITAMIN C IN SKIN

### VITAMIN C IN PHOTOAGEING

Vitamin C plays a potent role in diminishing the effects of free-radical damage, especially from ultraviolet light. The advantage of vitamin C over a sunscreen is that vitamin C can be absorbed into the cells and is generally still present about 30 to 36 hours after it has been applied topically to the skin.[8] It will, therefore, still give some sun protection despite sweating or swimming. Because of its antioxidant activity we can expect that vitamin C will tend to slow down photoageing while its DNA activity helps in reversing photodamage.

### VITAMIN C IN WRINKLES

Ascorbic acid is a potent agent to reduce wrinkles by promoting healthy collagen and elastin production in a lattice work. This improves tensile strength, thickening of the dermis, and the elastic

> *"Vitamin C seems to have limited activity on DNA itself but stimulates at least four genes related to collagen, and in particular, elastin production."*

layer just beneath the epidermis to give smoother younger-looking skin.[9, 10]

## VITAMIN C IN PIGMENTATION

Vitamin C affects pigmentation in two positive ways:

1. The creation of melanin is an oxidative process. A powerful antioxidant like vitamin C counteracts the oxidative process required to create melanin by binding to the reactive oxygen atoms that would drive pigment production, and as a result pigment production is limited.
2. Ascorbic acid also has a role as a tyrosinase inhibitor. It inhibits tyrosinase at two places in the metabolism of tyrosine into melanin. Tyrosinase is an enzyme essential for the formation of melanin, so if it is inhibited the production of pigmentation is reduced.[11, 12]

## VITAMIN C IN SCARS

If we deliver vitamin C in high dosage to the skin by iontophoresis, abnormal collagen that is often found under tethering scars may be replaced with normal collagen and the scars will fill up to a degree and become less noticeable.

Illustration shows the numerous effects of vitamin C when supplied in high dose (with retinol) by iontophoresis to a patient with severe chickenpox scars. You will notice that after 24 treatments of iontophoresis of vitamin C done over three months, the skin looks smoother, the pigmentation has been reduced, and the scars are flatter.

## USING VITAMIN C IN SKIN CARE

Vitamin C, as L-ascorbic acid, is commercially available as a dry white powder (technically called dehydroascorbic acid) which is relatively stable. When ascorbic acid powder is exposed to light and air, it slowly decomposes to oxidized ascorbic acid. When ascorbic acid crystals are mixed in water, the solution has an acidic pH. Under natural conditions the pH can easily be two or lower, depending on the concentration, meaning that this is quite a strong acid. Obviously the greater the saturation, the lower the pH and the more potent the acid, and the lower the pH, the more

*The effects of vitamin C can be enhanced by facilitating its penetration not merely through the stratum corneum but also into skin cells. These pictures demonstrate the value of employing iontophoresis in treating scars.*

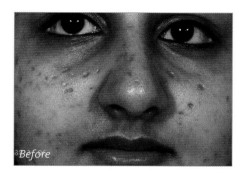
Before

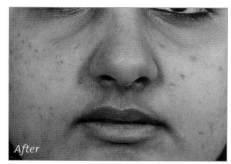
After

stable is the ascorbic acid solution. It is, however, much less stable than dry powdered ascorbic acid and rapidly decomposes to its oxidized form. It is thus strongly advisable to check the date of expiry on any ascorbic acid products. Products older than three weeks have to be 'preserved' in active form by other antioxidants such as ferulic acid.

Successful topical vitamin therapy for skin requires the freshest product that one can get. Under ideal conditions, one should use up the ascorbic acid product within three weeks of its date of manufacture. That is why the most reliable ascorbic acid products are sold unmixed and must be mixed just prior to use. The crystals are generally kept stable by filling the container with nitrogen instead of air to reduce the amount of oxygen that could degrade the vitamin C, as nitrogen is an inert gas that does not react with vitamin C.

With such precautions the solution, when finally made, ensures the freshest possible vitamin C. One really needs to have a concentration of ascorbic acid at a minimum of 10 per cent or more, to get the benefits of it. The pH will be about two, which will ensure the best penetration of the ascorbic acid through the horny layer of the skin. This pH is low and causes ascorbic acid to act like an alpha-hydroxy acid and 'softens the glue' between the cells of the horny layer of the epidermis. This increases the penetration of vitamin C into the deeper layers of the skin. That again, however, still does not ensure the best penetration of the cell. Vitamin C passes with difficulty into the cell wall because it is a water-soluble molecule, while the cell wall is mostly made up of lipid molecules which repel the water-soluble vitamin C. Once it is oxidized near the cell however, it can be taken into the cell itself by masquerading as glucose, and can be converted into active vitamin C by molecules called glutathione and alpha-lipoic acid.[13]

Ascorbic acid can be converted into more stable forms, but they tend to be more expensive. The various forms have different indications and we will try to explain how best to use them. If we combine L-ascorbic acid with another molecule that is selectively taken into the cell itself, we can solve the problem of stability as well as enhance the penetration through the skin and into the cell. An example of this form of vitamin C is magnesium (or sodium) ascorbyl phosphate, which is also water-soluble but is taken up into cells much more effectively. Once inside the cells, the vitamin C compound is easily converted to ascorbic acid, phosphate, and magnesium/sodium. Solutions of ascorbyl phosphates are more stable than conventional ascorbic acid and can last up to 200 days before there is any appreciable loss of activity. Lower concentrations, compared to plain ascorbic acid alone, are required to get the same amount of the acid into the cell.

Ascorbic acid and the ascorbyl salts like ascorbyl phosphate are both useful and provide similar results, but

> "My experience with this wonderful molecule makes me believe that we have entered a new era in vitamin C treatments for the skin."

have a different patient-usage profile. People with sensitive skins cannot use ascorbic acid, and so magnesium or sodium ascorbyl phosphate salts are preferred in active products. People with pigmentation problems should, however, avoid any product that peels the skin significantly. Ascorbic acid has an exfoliant property, so we generally recommend that patients with melasma (hormonal pigmentation) or other pigmentation problems should use products containing ascorbyl phosphate salts instead.

Both ascorbic acid and its salts are suitable for iontophoresis, and generally they can be used for professional treatments for pigmentation.[14] Ascorbic acid, with negative-current iontophoriesis becomes the anion ascorbate and is easily transported into cells and becomes available intracellularly as an enzymic co-factor or antioxidant. If patients complain that the ascorbic acid stings too much, it is better to use sodium ascorbyl phosphate for iontophoresis, as it has been shown to be more effective than magnesium ascorbyl phosphate.[14, 15]

There is a form of vitamin C that represents a much more elegant solution to get ascorbic acid inside cells at a high concentration. This method involves a lipid-soluble version of ascorbic acid. Ascorbic acid is water-soluble and is an antioxidant in the water phase, but it is largely extra-cellular because it is hard for it to pass through the naturally water-repellent cell walls. By attaching four palmitic acid molecules onto the ascorbic acid molecule we convert it into a lipid-soluble molecule that easily goes through the stratum corneum as well as through cell walls. This new molecule is called ascorbyl tetra-isopalmitate, which is the most stable version of vitamin C and has set the standards for vitamin C treatments.

The palmitic acid molecule is about the same weight as the ascorbic acid molecule, so vitamin C forms only about a fifth of this large molecule. However, with only a tiny amount of vitamin C, this fat-soluble form passes easily through the horny layer and enters the cell wall with great ease, allowing us to get up to ten times more active vitamin C into the cell itself compared to the other forms of vitamin C. As a result, there is more effective control of melanin formation, more action on DNA, greater collagen deposition, more elastin production, and more efficient intracellular antioxidant protection.

My experience with this wonderful molecule makes me believe that we have entered a new era in vitamin C treatments for the skin. A drawback of this molecule, however, is that this vitamin C is fat-soluble and thus cannot be used for iontophoresis or sonophoresis. It is also, paradoxically, not an antioxidant outside cells and only becomes one as soon as the four palmitic acid molecules have been removed in the cell cytoplasm. It is clinically highly effective. We have combined it with vitamin A and seen rapid smoothing without any irritation and excellent lightening of the skin. If it is also combined with a wide antioxidant brigade and effective UVA protection, then pigmented marks seem to fade away in some people.

Ascorbyl tetra-isopalmitate should not to be confused with ascorbyl palmitate, or ascorbyl di-palmitate, which has been used in low doses by many cosmetics manufacturing companies in the past. These molecules are far less fat-soluble than the ascorbyl tetra-isopalmitate. Ascorbyl palmitate has the advantage of being somewhat fat-soluble, but unfortunately, it is chemically difficult to include in cosmetic formulations in high doses. For high doses, alcohol has to be employed, and that limits its use severely. The fact remains that not nearly as much vitamin C is achieved inside the cell as with the ascorbyle tetra-isopalmitate.

## IN SUMMARY, THIS IS HOW TO USE THE VARIOUS FORMS OF VITAMIN C

**I. ASCORBIC ACID:** should be used as fresh as possible. Use this on people with tough, rough and wrinkled skin, but avoid using it on people prone to acne because the exfoliation can initially aggravate the acne. It can be used for home vitamin C peels with great success. Ascorbic acid is excellent for iontophoresis and can be used for sonophoresis.

**II. MAGNESIUM (SODIUM) ASCORBYL PHOSPHATE**: use as fresh as possible and discard anything that is older than a year after production. Use this on normal, sensitive, or pigmented problem skin. It can be used on acne but may initially aggravate it. Sodium ascorbyl phosphate is the best for iontophoresis, and both magnesium and sodium ascorbyl phosphate are suitable for sonophoresis.

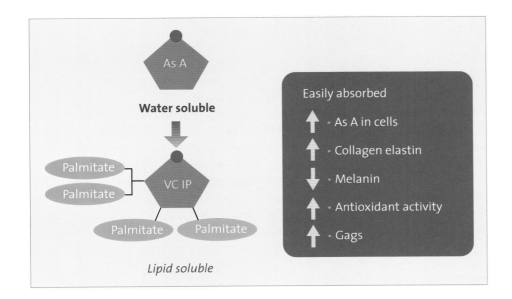

*Diagram illustrating the conversion of water-soluble ascorbic acid to lipid-soluble by "wrapping' the molecule in four palmitic acid molecules.*

**III. ASCORBYL TETRA-ISOPALMITATE:** has excellent stability, but for the best results the product should not be older than 24 months. It gives the best levels of vitamin C inside the cells and can be used on delicate skins at high doses. Use it in preference to any other forms for pigmentation, wrinkles, and even acne.

In time we will see more effective forms of vitamin C because researchers are working hard to find the ideal molecule. They have also combined it with vitamin A to try and get the benefits of both of these wonderful molecules. Ultimately whatever molecule is developed, it will always deliver L-ascorbic acid into the cell because that is where we need this irreplaceable vitamin for healthy skin.

## VITAMIN E

Vitamin E is fat-soluble and is one of the most powerful antioxidants protecting the lipid phases of our body, such as the membranes of our cells, which are composed largely of fatty molecules called lipoproteins; vitamin E is found among those lipoprotein molecules.

There are eight variants of vitamin E: four tocopherols (tocopherol is the chemical name for vitamin E) and four tocotrienols. The various family members of the vitamin E household are named as alpha-, beta-, gamma- and delta-tocopherol and the four variants of the tocotrienols: alpha, beta, gamma and delta variants according to their unique chemical structures. The ones that we most commonly encounter are alpha-tocopherol and gamma-tocopherol.

> "We need vitamin E in our skin to help us protect the cells from the damaging effects of free radicals, particularly those induced by UVA rays."

Vitamin E is unusual amongst natural chiral molecules. 'Chiral' is a concept in chemistry that means a molecule has such a structure that it cannot be superimposed on its mirror image. In simple terms, certain molecules will fit into receptors while the same molecule that is turned to the opposite side will not fit into the same receptors. When molecules with opposite turns are mixed in a single formula it will often result in the deactivation of the intended effect of the formula. In other words it doesn't work.

Whereas most others are in the 'L-' form like L-ascorbic acid or 'left turned', tocopherol is naturally active as D-alpha tocopherol, or 'right-turned'. Vitamin E. Synthetic vitamin E is a 50:50 mixture of D-tocopherol and L-tocopherol and is designated as D-L-tocopherol. Synthetic vitamin E can be 'filtered' to isolate the naturally identical D-tocopherol, but of course, this makes it much more expensive.

The most common sources for vitamin E in our food are vegetable oils, nuts, green leafy vegetables, and fortified cereals. It seems as though humans need all the variants of tocopherol and tocotrienol.[16] While vitamin E appears to be mainly an antioxidant, it also has some metabolic activity.[17] In this capacity it may assist in preventing DNA mutations and cancers,[18] but all studies show no effects from only using one version of vitamin E. Vitamin E has been shown to lengthen the life of cells by delaying shortening of the telomeres.[19] Tocotrienols seem to protect nerve tissue.[20] We should always eat diets and take supplements that give us a wide intake of all the different tocopherols and tocotrienols.

Once vitamin E has been deactivated by interacting with a lipid-phase free radical, it can be reactivated by vitamin C, which is surprising because vitamin C is a water-phase free-radical scavenger. Vitamin E can also be reactivated by retinol, ubiqionol, coenzyme Q10, alpha lipoic acid, and glutathione.[21, 22] These are other antioxidants normally present in skin cells. This reactivation process explains why vitamin E deficiency is not as common as one would think, as many people have a less than adequate dietary intake of vitamin E.[23] Vitamin E also has metabolic effects such as inhibiting certain cell signalling molecules (e.g. protein kinase –C).[24]

We need vitamin E in our skin to help us protect the cells from the damaging effects of free radicals, particularly those induced by UVA rays.[25, 26] Those free radicals need to be deactivated as rapidly as possible, otherwise the damage to the cellular membranes results in defects in the membrane and typically 'sunburn cells' develop immediately prior to the death of such damaged cells. If one has sufficient vitamin E in the skin, it limits the number of 'sunburn cells' formed by too much UV exposure.[27]

There are therefore numerous reasons that I believe that we should keep our skin well supplied with vitamin E at all times. Although vitamin E does not have an established function in changing the hydration of skin, topical vitamin E preparations do tend to improve skin hydration, probably in an indirect manner. At present we still know very little more for certain about vitamin E, but one suspects that there is more to it than is described here.

## VITAMIN B3 (NICOTINIC ACID, NICOTINAMIDE, NIACIN)

Niacin and nicotinamide are both called Vitamin B3 (which is an important part of the water-soluble vitamin B group) but nicotinamide adenine dinucleotide (NAD) is the bio-active form of niacin, and NAD is involved particularly in energy metabolism.

When people are chronically deficient in vitamin B3 they develop a condition called pellagra with dark blotchy pigmented skin. Nicotinic acid is transported through skin with difficulty because it is water-soluble and has to be specially formulated to allow it to penetrate. Once inside the cells, it is converted into NAD which:

(A) facilitates repair of DNA, which is a very important step in preventing photoageing, [28]
(B) enhances epidermal growth and repair. [29] This would work in parallel with vitamin A;
(C) controls the release of melanin from the melanocytes to the surrounding keratinocytes. This has important value for skin in reducing excess or abnormal pigmentation; [30]
(D) acts as a natural sunscreen and thus further limits DNA damage.[31]

*This patient illustrates the benefits that came in three months from using niacinamide with undecylenoyl phenylalanine in conjunction with retinyl palmitate.*

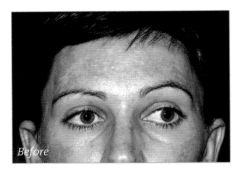
Before

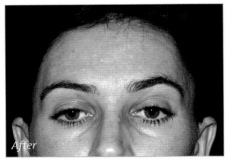
After

### VITAMIN B5 PANTHENOL

Vitamin B5, pantothenic acid, and pre-cursor panthenol, is another water-soluble B vitamin. Virtually all our foods contain it, but in low doses. Higher doses are found in whole grain cereals, beans, eggs and meat. Vitamin B5 is important in the production of co-enzyme A and the production of carbohydrates, fats, and proteins. It has important effects such as increasing the production of glutathione, one of our most important intracellular antioxidants, and these combined effects protect the mitochondria and other intracellular organelles.[32]

Vitamin B5, similar to vitamin E, is naturally active only as a D-isomer: D-panthenol. The L-panthenol may actually reverse the effects of D-panthenol, and so D-L-panthenol should never be used. One advantage of panthenol in skin care is that it is easily absorbed into skin and skin cells and so soothes the skin and increases the hydration. Panthenol is one of the safest ingredients one can apply to skin.[33]

### VITAMIN B12 (METHYLCOBALAMIN)

This is the best-known member of the B vitamin group. It is an essential co-enzyme, and few people realize that it is extremely sensitive to light, and as a result we easily deplete our levels of methylcobalmin when we are out in the sun. It is an important nutrient for every cell of the body and is involved in DNA synthesis, energy production, and red blood cell formation.[34] A deficiency of vitamin B12 causes a serious red blood cell anaemia called pernicious anaemia. Vitamin B12 is important in the maintenance of nerve health and is an essential way to reduce the production of homocysteine in the body.[35] Topically on the skin, vitamin B12 plays an important part in keeping cells as healthy as we can make them by promoting healthy DNA.

Vitamin B12 (methylcobalamin) is classed amongst the water-soluble vitamins, though it is not easily soluble in water. That may explain why it is not excreted as rapidly as the other water-soluble vitamins and is stored in the liver and kidneys. We get our vitamin B12 only through animal products like liver and meat, fish, chicken, eggs and milk. Strangely enough, animals cannot make vitamin B12 but rely on bacteria in their bowels to synthesise it. Vitamin B12 is absorbed through the bowel but we need a so-called 'intrinsic factor' for its absorption. Intrinsic factor is produced in the stomach and as we get older we produce less intrinsic factor and as a result we absorb less vitamin B12.

Vitamin B12 can be absorbed through the mouth, and in this case intrinsic factor is not required. However, the vitamin has to be kept in the mouth as long as possible to ensure absorption. If you swallow it, vitamin B12 can only be absorbed with intrinsic factor. Most supplements do not supply adequate doses of vitamin B12, even though we only need very tiny amounts of it each day. Cyanocobalamin is often used as a vitamin B12 supplement, but is never found in nature, so it is preferable to use nature-identical methylcobalamin. To keep tissues as healthy as possible, vitamin B12 should be supplemented daily. Homocysteine is a toxin that accumulates in our body and vitamin B12 is important in keeping homocysteine levels low. As the name implies, cobalt is an essential part of methylcobalamin, which acts as a co-factor in metabolism and also in the formation of DNA. It turns out that vitamin B12 is sensitive to light, and we need to supplement the skin with topical vitamin B12 to make up for the loss incurred every time we go out into sunlight. The demand is not as great as the need for vitamin A, but for ideal function of skin we need to supplement vitamin B12 regularly.

## VITAMIN D

A steroid type of vitamin related to cholesterol, vitamin D is formed in the skin only on exposure to UVB in sunlight. We naturally have in our skin the pre-cursor of vitamin D, which is a chromophore for UVB, and on exposure to UVB light this molecule undergoes changes and in about 20 minutes becomes active vitamin D3. (There are three types of vitamin D: D1, D2 and D3. Vitamin D3 is the most effective of all three forms.)

We cannot make vitamin D in the early part of the morning or in the late afternoon, but make it preferentially from about 11:00 till about 15:00 because UVB rays are filtered out by the sunlight coming obliquely through the atmosphere before or after that time. People living in the tropics make vitamin D during winter because the sun's rays close to the equator are almost directly overhead and do not have to pass very obliquely through the atmosphere, and so UVB rays reach the surface in the midday period. From about mid-autumn through to mid-spring one cannot make sufficient vitamin D in temperate areas because there are no UVB rays. People living in very northern or very southern areas, like northern America, Europe, Siberia, or South Island, New Zealand, make modest amounts of vitamin D only in summer because that is the only time that UVB rays reach the Earth in those areas.[36]

On the other hand, UVA rays pass through the atmosphere with minimal interference for most of the day, and that is important because UVA rays destroy vitamin D![37] That also means

*"One advantage of panthenol in skin care is that it is easily absorbed into skin and skin cells and so soothes the skin and increases the hydration."*

> "It is estimated that 50 per cent of the population in Europe, the USA, and Asia are deficient in Vitamin D."

that vitamin D is a natural sunscreen: it absorbs UVB rays (to manufacture vitamin D) as well as UVA and is then destroyed. This destruction of vitamin D may explain why we never get vitamin D toxicity even after intense and prolonged exposure to sunlight when we may make tens of thousands of international units of vitamin D.

It is estimated that 50 per cent of the population in Europe, the USA, and Asia are deficient in Vitamin D. This deficiency is more harmful than once believed, because we are learning that vitamin D deficiency not only increases the risk of osteoporosis and bone fractures, but also the risk of muscle weakness, depression, obesity, and many cancers. Natural selection may have favoured the development of lighter skin colour in northern latitudes, because the lighter the skin colour, the easier it is to make vitamin D. People with dark skins can suffer from vitamin D deficiency (rickets) if they live in very northern countries, whereas in the tropics people with dark coloured skin easily make vitamin D and at the same time are more protected from deeper sun damage.

There is concern that most people make less than optimal levels of vitamin D, not only because of seasonal variations in UVB irradiation, but also because of avoiding sunlight to minimize photodamage, abnormal pigmentation, and cancer. The skin of over 60 per cent of people in the world has a natural SPF value greater than three, and that interferes with natural vitamin D production. Sunblocks may also dramatically reduce vitamin D production.[38] One cannot rely on food as the only source because the levels of vitamin D are insufficient. That means that vitamin D supplementation assumes a greater importance to maintain calcium levels and healthy bones and teeth and reduce the risk of osteoporosis and cancer in later life.[39]

The view of all the researchers concerned with vitamin D is that all women should supplement with vitamin D from their early twenties to strengthen bones. In addition to maintaining healthy bones and teeth, Vitamin D also helps relieve depression, protects against heart disease, and has anticarcinogenic effects. Research published in August 2007 suggested that many hundreds of thousands of cases of cancer in the world can be prevented every year through regular supplementation with vitamin D.[40, 41]

Vitamin D makes us optimistic, boosts our immunity, and is also useful in treating some types of psoriasis. There is growing evidence that it helps protect us against multiple sclerosis, lupus erythematosus, and fibromyalgia. It has also been shown to reduce insulin resistance in diabetics. We should never forget that vitamin D shares receptors for many genes with vitamin A, and when levels of vitamin A and D are ideal, our cells function much better and it is easier to keep skin healthy.

## PRECAUTIONS, SIDE-EFFECTS, AND INTERACTIONS

Dosages of vitamin D up to 2400 IU/day rarely cause adverse reactions.

Continuous dosage above 3800 IU/day may cause excessive calcium levels in the blood, which can result in nausea, vomiting, weakness, headache, sleepiness, dry mouth, constipation, a metallic taste in the mouth, and muscle and bone pain. If left untreated this can progress to pancreatitis, light sensitivity, runny nose, itchy skin, elevated body temperature, decreased libido, elevated liver enzymes, kidney stones, high blood pressure, and an irregular heartbeat.

We have learned that we have to pay much more attention to vitamin D, which is well described as the mood enhancing, bone strengthening, cancer-protecting vitamin which also helps to keep skin healthy. Supplement with 2000 IU of vitamin D3 every day in summer and winter, because we are generally indoors at the times when we could make vitamin D naturally. On weekends people may make less than ideal amounts of vitamin D because they use sunscreens.

## CAROTENOIDS

Carotenoids like beta-carotene are a precursor of vitamin A. The carotenoid chemicals are yellow, orange, or red and are almost universally distributed in living tissues. When we ingest certain carotenes, we are able to synthesise vitamin A in our bodies. Beta-carotene is one of the safest forms of vitamin A. We are able to make as much vitamin A as we need from beta-carotene without the risk of vitamin A toxicity.

There are also non-vitamin-A carotenoids such as lycopene, lutein, and zeaxanthin that are essential for skin and add to the protection from UV light.[42] Lutein specifically helps to absorb blue light.[43, 44] The important role for carotenoids is as an antioxidant,[45] as they are among of the most effective antioxidants and are relatively stable. One molecule of beta-carotene can counteract a vast number of free radicals. Beta-carotene is an important defence against infra-red rays that produce a huge array of free radicals.[46] Alpha-carotene has similar effects to beta-carotene. Asians have a higher level of beta-carotene in their skin naturally, which may be the reason that their skin appears more yellow and also looks younger.

The controversy about whether beta-carotene promotes lung cancer concerns only its use when taken orally.[47, 48] When applied to the skin, beta-carotene is placed directly where it is needed and is constantly used up. The antioxidant vitamins seem to be intertwined in their action and they should be used together to get their best effects. Do not make the mistake of thinking that there is one particular antioxidant that is superior to all others; they work in concert under natural conditions.

## COENZYME Q10

This molecule is part of the natural antioxidant network. We naturally make coenzyme Q10 but not in ideal doses. This antioxidant is especially useful in the mitochondria, which are the powerhouses of the cell. During the production of energy in the mitochondria, free radicals are

"*Vitamin D shares receptors for many genes with vitamin A, and when levels of vitamin A and D are ideal, our cells function much better and it is easier to keep skin healthy.*"

generated in great quantities at the end of the energy cycle. Since this occurs in the area immediately next to the mitochondrial DNA, these free radicals can be dangerous because they can damage the DNA directly.

Coenzyme Q10 is similar to idebenone, a synthetic molecule that is a very powerful antioxidant specifically for the conditions in the mitochondria. The advantage of coenzyme Q10 is that it is readily metabolized into a form that interacts easily with the mitochondria. In addition, coenzyme Q10 is valuable in cells, as it converts inactive vitamin C radical back into the active antioxidant form.

## ALPHA-LIPOIC ACID

This amazing molecule is normally found in cells and can be manufactured in the body. We rely on our nutrition to boost the levels in our body and we find it in liver, red meat, spinach, broccoli, and red potatoes.

Alpha-lipoic acid has a wide variety of roles and is a very powerful antioxidant in both the lipid and the water phases of the cells. It also has a specific role in utilizing oxygen, normalizing liver metabolism, reversing nerve damage in diabetes, and slowing down ageing of the brain,[49] and even has shown strong anti-cancerous activity. Because alpha-lipoic acid can recycle vitamin E from an oxidized radical back to the active antioxidant, it prevents vitamin E deficiency.[23] Lester Packer considers alpha-lipoic acid to be one of the most important nutrients we need to protect our mitochondria.[50, 51]

## GLUTATHIONE

This is another antioxidant that is made in the body and is one of the most pervasive and important ones. It is interesting to note that only glutathione among all of the skin antioxidants is not depleted by exposure to UV light. It is a truly potent antioxidant that cannot be supplied by mouth or applied to the skin because it is decidedly unstable outside the cell.

## THE ANTIOXIDANT NETWORK

Vitamins C and E, alpha-lipoic acid, coenzyme Q10, and glutathione form what has been described as the Antioxidant Network, because they are able to recycle each other.[22] In general if any two are supplied, the levels of the other members of the antioxidant network will automatically be built up.[22] An example of this is that animals fed with vitamins C and alpha-lipoic acid will never become vitamin-E deficient even when they are deprived of vitamin E.[23] This recycling becomes very convenient when formulating cosmetic creams, because alpha-lipoic acid is an unpleasant smelling molecule to formulate into sophisticated skin care. It is supposed to be effective in both water and lipid phases, yet it is only really soluble in alcohol, which unfortunately means that the product will not only have an offensive smell at adequate doses but will also be an irritant because of the alcohol. By supplying vitamins E and C the levels of alpha-lipoic acid are automatically boosted.

There is a similar disadvantage for coenzyme Q10; however, it has nothing to do with smell. It is just too expensive

> "Vitamins C and E, alpha-lipoic acid, coenzyme Q10, and glutathione form what has been described as the Antioxidant Network, because they are able to recycle each other."

to incorporate into products without inflating the cost of ethically priced creams. Fortunately, the networking effect among the antioxidants means that we can achieve good levels of the alpha-lipoic acid and coenzyme Q10 by making sure that the skin-care product contains good doses of vitamins C, E, and other antioxidants such as beta-carotene. In fact, by following this course of action one is able to build up glutathione, one of the most powerful intra-cellular antioxidants, which cannot be supplemented either topically or orally.

## SELENIUM

Selenium is an element that functions as a coenzyme in a chemical process important for removing free radicals. It has been shown that skin cancer is more common in people who have a low dietary intake of selenium.[52] It is not easily absorbed by skin when applied topically and is generally used as a dietary supplement. Only trace amounts of selenium are required as supplementation, and too much may be harmful. It is interesting that New Zealanders are particularly susceptible to selenium deficiency, as selenium is present only at low levels in their soil and thus levels are low in much of their food. When such localized circumstances are prevalent, it is important to correct those by supplementation.[39]

## SUPEROXIDE DISMUTASE

This is an enzyme that removes strong oxygen radicals and converts them into oxygen and hydrogen peroxide, which eventually breaks down into water and singlet oxygen. This is a powerful and common antioxidant found in tissues that confront oxygen, and various forms are known depending on which metal they are associated with, for example, zinc, manganese, or iron. This is a complex metalloprotein that cannot easily pass into the skin.

## ZINC

The element zinc is a powerful antioxidant agent that protects against UV radiation, enhances wound healing, contributes to immune and neural functions, and decreases the relative risk of cancer and cardiovascular disease.[53] All body tissues contain zinc, but in skin it is five to six times more concentrated in the epidermis than the dermis.[54] Topical zinc ions may provide an important and helpful antioxidant defence for skin.[55]

## REFERENCE

1. Kagan, V., et al., *Ultraviolet light-induced generation of vitamin E radicals and their recycling. A possible photosensitizing effect of vitamin E in skin.* Free Radic Res Commun, 1992. **16**(1): p. 51-64.
2. Jimenez, S.A., et al., *Increased collagen biosynthesis and increased expression of type I and type III procollagen genes in tight skin (TSK) mouse fibroblasts.* J Biol Chem, 1986. **261**(2): p. 657-62.
3. Geesin, J.C., et al., *Ascorbic acid specifically increases type I and type III procollagen messenger RNA levels in human skin fibroblast.* J Invest Dermatol, 1988. **90**(4): p. 420-4.
4. Nusgens, B.V., et al., *Topically applied vitamin C enhances the mRNA level of collagens I and III, their processing enzymes and tissue inhibitor of matrix metalloproteinase 1 in the human dermis.* J Invest Dermatol, 2001. **116**(6): p. 853-9.

5. Catani, M.V., et al., *Biological role of vitamin C in keratinocytes.* Nutr Rev, 2005. **63**(3): p. 81-90.
6. Davidson, J.M., et al., *Ascorbate differentially regulates elastin and collagen biosynthesis in vascular smooth muscle cells and skin fibroblasts by pretranslational mechanisms.* J Biol Chem, 1997. **272**(1): p. 345-52.
7. Duarte, T.L., M.S. Cooke, and G.D. Jones, *Gene expression profiling reveals new protective roles for vitamin C in human skin cells.* Free Radic Biol Med, 2009. **46**(1): p. 78-87.
8. Burke, K.E., *Interaction of vitamins C and E as better cosmeceuticals.* Dermatol Ther, 2007. **20**(5): p. 314-21.
9. Pinnell, S.R., *Cutaneous photodamage, oxidative stress, and topical antioxidant protection.* J Am Acad Dermatol, 2003. **48**(1): p. 1-19; quiz 20-2.
10. Fitzpatrick, R.E. and E.F. Rostan, *Double-blind, half-face study comparing topical vitamin C and vehicle for rejuvenation of photodamage.* Dermatol Surg, 2002. **28**(3): p. 231-6.
11. Matsuda, S., et al., *Inhibitory effects of a novel ascorbic derivative, disodium isostearyl 2-O-L-ascorbyl phosphate on melanogenesis.* Chem Pharm Bull (Tokyo), 2008. **56**(3): p. 292-7.
12. Farris, P.K., *Topical vitamin C: a useful agent for treating photoageing and other dermatologic conditions.* Dermatol Surg, 2005. **31**(7 Pt 2): p. 814-7; discussion 818.
13. Savini, I., S. Duflot, and L. Avigliano, *Dehydroascorbic acid uptake in a human keratinocyte cell line (HaCaT) is glutathione-independent.* Biochem J, 2000. **345 Pt 3**: p. 665-72.
14. Huh, C.H., et al., *A randomized, double-blind, placebo-controlled trial of vitamin C iontophoresis in melasma.* Dermatology, 2003. **206**(4): p. 316-20.
15. Marra, F., et al., *In vitro evaluation of the effect of electrotreatment on skin permeability.* J Cosmet Dermatol, 2008. **7**(2): p. 105-11.
16. Packer, L., S.U. Weber, and G. Rimbach, *Molecular aspects of alpha-tocotrienol antioxidant action and cell signalling.* J Nutr, 2001. **131**(2): p. 369S-73S.
17. Tucker, J.M. and D.M. Townsend, *Alpha-tocopherol: roles in prevention and therapy of human disease.* Biomed Pharmacother, 2005. **59**(7): p. 380-7.
18. Foote, J.A., et al., *Chemoprevention of human actinic keratoses by topical DL-alpha-tocopherol.* Cancer Prev Res (Phila), 2009. **2**(4): p. 394-400.
19. Tanaka, Y., Y. Moritoh, and N. Miwa, *Age-dependent telomere-shortening is repressed by phosphorylated alpha-tocopherol together with cellular longevity and intracellular oxidative-stress reduction in human brain microvascular endotheliocytes.* J Cell Biochem, 2007. **102**(3): p. 689-703.
20. Bourre, J.M., *Effects of nutrients (in food) on the structure and function of the nervous system: update on dietary requirements for brain. Part 1: micronutrients.* J Nutr Health Ageing, 2006. **10**(5): p. 377-85.
21. Sies, H. and W. Stahl, *Vitamins E and C, beta-carotene, and other carotenoids as antioxidants.* Am J Clin Nutr, 1995. **62**(6 Suppl): p. 1315S-1321S.
22. Thiele, J.J., et al., *The antioxidant network of the stratum corneum.* Curr Probl Dermatol, 2001. **29**: p. 26-42.
23. Podda, M., et al., *Alpha-lipoic acid supplementation prevents symptoms of vitamin E deficiency.* Biochem Biophys Res Commun, 1994. **204**(1): p. 98-104.
24. Singh, U., S. Devaraj, and I. Jialal, *Vitamin E, oxidative stress, and inflammation.* Annual review of nutrition, 2005. **25**: p. 151-74.
25. Foote, J.A., et al., *Chemoprevention of human actinic keratoses by topical DL-alpha-tocopherol.* Cancer Prev Res (Phila Pa), 2009. **2**(4): p. 394-400.
26. Sakagami, H., et al., *Effect of alpha-tocopherol on cytotoxicity induced by UV irradiation and antioxidants.* Anticancer Res, 1997. **17**(3C): p. 2079-82.
27. Lin, J.Y., et al., *UV photoprotection by combination topical antioxidants vitamin C and vitamin E.* J Am Acad Dermatol, 2003. **48**(6): p. 866-74.
28. Benavente, C.A., M.K. Jacobson, and E.L. Jacobson, *NAD in skin: therapeutic approaches for niacin.* Curr Pharm Des, 2009. **15**(1): p. 29-38.
29. Jacobson, E.L., et al., *A topical lipophilic niacin derivative increases NAD, epidermal differentiation and barrier function in photodamaged skin.* Exp Dermatol, 2007. **16**(6): p. 490-9.

30. Hakozaki, T., et al., *The effect of niacinamide on reducing cutaneous pigmentation and suppression of melanosome transfer.* Br J Dermatol, 2002. **147**(1): p. 20-31.
31. Damian, D.L., *Photoprotective effects of nicotinamide.* Photochemical & photobiological sciences : Official journal of the European Photochemistry Association and the European Society for Photobiology, 2010. **9**(4): p. 578-85.
32. Slyshenkov, V.S., D. Dymkowska, and L. Wojtczak, *Pantothenic acid and pantothenol increase biosynthesis of glutathione by boosting cell energetics.* FEBS letters, 2004. **569**(1-3): p. 169-72.
33. Bissett, D.L., *Common cosmeceuticals.* Clin Dermatol, 2009. **27**(5): p. 435-445.
34. Zempleni, J., *Handbook of vitamins.* 4th ed2007, Boca Raton: Taylor & Francis. xii, 593 p.
35. Seal, E.C., et al., *A randomized, double-blind, placebo-controlled study of oral vitamin B12 supplementation in older patients with subnormal or borderline serum vitamin B12 concentrations.* J Am Geriatr Soc, 2002. **50**(1): p. 146-51.
36. Webb, A.R., L. Kline, and M.F. Holick, *Influence of season and latitude on the cutaneous synthesis of vitamin D3: exposure to winter sunlight in Boston and Edmonton will not promote vitamin D3 synthesis in human skin.* J Clin Endocrinol Metab, 1988. **67**(2): p. 373-8.
37. Webb, A.R., B.R. DeCosta, and M.F. Holick, *Sunlight regulates the cutaneous production of vitamin D3 by causing its photodegradation.* J Clin Endocrinol Metab, 1989. **68**(5): p. 882-7.
38. Holick, M.F., *Sunlight, UV-radiation, vitamin D and skin cancer: how much sunlight do we need?* Adv Exp Med Biol, 2008. **624**: p. 1-15.
39. Nakamura, K., *Vitamin D and prevention of osteoporosis: Japanese perspective.* Environ Health Prev Med, 2006. **11**: p. 271-276.
40. Garland, C.F., et al., *What is the dose-response relationship between vitamin D and cancer risk?* Nutr Rev, 2007. **65**(8 Pt 2): p. S91-5.
41. Garland, C.F., et al., *Vitamin D for cancer prevention: global perspective.* Ann Epidemiol, 2009. **19**(7): p. 468-83.
42. Stahl, W. and H. Sies, *Carotenoids and protection against solar UV radiation.* Skin Pharmacol Appl Skin Physiol, 2002. **15**(5): p. 291-6.
43. Nilsson, S.E., et al., *Ageing of cultured retinal pigment epithelial cells: oxidative reactions, lipofuscin formation and blue light damage.* Doc Ophthalmol, 2003. **106**(1): p. 13-6.
44. Sies, H. and W. Stahl, *Non-nutritive bioactive constituents of plants: lycopene, lutein and zeaxanthin.* Int J Vitam Nutr Res, 2003. **73**(2): p. 95-100.
45. Krinsky, N.I., *Antioxidant functions of carotenoids.* Free Radic Biol Med, 1989. **7**(6): p. 617-35.
46. Darvin, M.E., et al., *Topical beta-carotene protects against infra-red-light-induced free radicals.* Experimental dermatology, 2011. **20**(2): p. 125-9.
47. Albanes, D., et al., *Alpha-Tocopherol and beta-carotene supplements and lung cancer incidence in the alpha-tocopherol, beta-carotene cancer prevention study: effects of base-line characteristics and study compliance.* J Natl Cancer Inst, 1996. **88**(21): p. 1560-70.
48. Holick, C.N., et al., *Dietary carotenoids, serum beta-carotene, and retinol and risk of lung cancer in the alpha-tocopherol, beta-carotene cohort study.* Am J Epidemiol, 2002. **156**(6): p. 536-47.
49. Hagen, T.M., et al., *Feeding acetyl-L-carnitine and lipoic acid to old rats significantly improves metabolic function while decreasing oxidative stress.* Proc Natl Acad Sci U S A, 2002. **99**(4): p. 1870-5.
50. Packer, L., E.H. Witt, and H.J. Tritschler, *Alpha-lipoic acid as a biological antioxidant.* Free Radic Biol Med, 1995. **19**(2): p. 227-50.
51. Packer, L., *Alpha-Lipoic acid: a metabolic antioxidant which regulates NF-kappa B signal transduction and protects against oxidative injury.* Drug Metab Rev, 1998. **30**(2): p. 245-75.
52. Knekt, P., et al., *Serum selenium and subsequent risk of cancer among Finnish men and women.* J Natl Cancer Inst, 1990. **82**(10): p. 864-8.
53. Ames, B.N., *A role for supplements in optimizing health: the metabolic tune-up.* Arch Biochem Biophys, 2004. **423**(1): p. 227-34.
54. Rostan, E.F., et al., *Evidence supporting zinc as an important antioxidant for skin.* Int J Dermatol, 2002. **41**(9): p. 606-11.
55. Chan, S., B. Gerson, and S. Subramaniam, *The role of copper, molybdenum, selenium, and zinc in nutrition and health.* Clin Lab Med, 1998. **18**(4): p. 673-85.

# Chapter 12
# OTHER IMPORTANT MOLECULES FOR PROTECTING SKIN

### RESVERATROL

This is an exciting molecule that has become an essential part of an anti-ageing program for skin and the body. Resveratrol exists as both *cis-* and *trans-*isomers but mainly as *trans-*resveratrol, which is the biologically active isomer.[1] '*Cis-* and *trans-*isomers' means that the chemical composition of the molecules is identical, but certain atoms in critical positions sit in opposite places.

Resveratrol has been shown to be a cancer preventive agent as a result of the expression of various growth factors,[2, 3] some of which are expressed by keratinocytes. Topical application of resveratrol has been shown to hasten wound repair with greater epithelial growth, more collagen, more elastin, and improved histological architecture of the skin.[4]

Resveratrol lightens the skin colour and so is valuable in treating pigmentation problems. It exerts its effect on the melanocyte by reducing the activity of genes that promote the formation of the enzyme tyrosinase [5] and protects skin from UVA damage resulting in a reduction of the skin's ability to tan.[6] Because resveratrol additionally protects against UVB damage, it clearly has broad powers in reducing UV damage in general.[7]

The heme oxygenase-1 gene is rapidly induced by UVA, and antioxidant vitamins C and E or beta-carotene do not change this condition. On the other hand, resveratrol rapidly causes the degradation of a molecule called heme oxygenase (iron-protoporphyrin IX), a pro-oxidant (like a free radical) that damages nerves, thus resveratrol is believed to protect nerve tissue.[8, 9] Resveratrol also 'rejuvenates' the DNA of cells[10] and protects it from free-radical damage.[11]

## OTHER PLANT-DERIVED ANTIOXIDANTS

There are very many plant-derived antioxidants, but the following molecules are derived from plants that grow exclusively in South Africa. These molecules have been shown to be unusually powerful antioxidants.

## GREEN ROOIBOS TEA EXTRACT (ASPALANTHUS LINEARIS LEAF EXTRACT)

The rooibos bush is particular to South Africa in the arid west coast regions. It was used by the San people and soon gained popularity amongst the immigrant Dutch settlers. The unfermented leaves contain very little tannin and no caffeine and contain a unique polyphenol called aspalanthin, along with many other powerful molecules such as flavonoids and flavones such as rutin, luteolin, and quercetin[12]. These act like vitamin C but are more active, and they do not become inactive due to exposure to light.

Rooibos extract also contains iron, magnesium, potassium, fluoride, manganese, zinc, and calcium. Even though only tiny amounts of rooibos extracts pass through the skin barrier they still have decisive and powerful effects. Presumably because of its strong antioxidant activity, rooibos extract has a role in decreasing melanin formation, as has been seen in clinical studies. Research suggests that rooibos may promote healthy and 'younger-acting' chromosomes.[13]

## ORGANIC HONEYBUSH TEA EXTRACT (CYCLOPIA GENISTOIDES LEAF EXTRACT)

This is a virtually caffeine- and tannin-free tea, the antioxidant activity of which exceeds that of rooibos extract and is close to that of green tea. It has major phenolic compounds: mangiferin and xanthone as well as many others.[14] Polyphenols are increasingly recognised as some of the most effective and highly active antioxidant molecules available.

Honeybush extract is also rich in superoxide dismutase and contains numerous minerals like calcium, copper, zinc, magnesium, manganese, and sodium. Besides being an antioxidant, honeybush also has an anti-cancer-proliferative activity.[15] The combination of honeybush extract with rooibos extract may well be an unusually powerful way of keeping cells healthy.[16, 17]

I believe that green tea works in a complementary fashion with the two South African tea extracts in making healthy skin as well as providing healthy bodies. Combining these three agents provides another part of the wonderful antioxidant brigade, which complements the actions of vitamin C and vitamin E.

## GREEN TEA EXTRACT WITH EPIGALLOCATECHIN GALLATE (EGCG)

Unfermented tealeaves contain epigallocathechin gallate (EGCG), a potent singlet-oxygen antioxidant. It prevents lipid peroxidation and restores epidermal glutathione and glutathione peroxidase, which are such important antioxidants within the cell and cannot be supplemented by mouth or by topical applications.

One of the amazing powers of ECGC is that it seems to be anticarcinogenic.[18] With topical application it has been shown to inhibit ultraviolet- induced cancers.[18] When we check the DNA we find that it protects against UVA-induced DNA damage as demonstrated by reduced formation of molecules known as pyrimidine dimers. These so-called dimers are base pairs of the DNA that become linked in the process of damage and lead to skin cancer.

ECGC has a specific effect on cells that have been damaged by sunlight and can resuscitate damaged cells;[19] furthermore, it protects Langerhans cells from UV irradiation. ECGC is in fact somewhat like a natural sunscreen, yet has no sunscreen activity in the conventional sense. In spite of this, it inhibits the normal erythema response to UVB irradiation.[19] There is some evidence to suggest that it might specifically protect skin from developing melanoma.[20] This may be one reason that melanoma is uncommon amongst Japanese. ECGC stimulates proliferation of keratinocytes that causes a thicker stratum spinosum.[21] It also protects fibroblasts from UVA damage and reduces the activity of the collagenases (enzymes which break down collagen) that would destroy collagen.[22, 23] ECGC has antibacterial and bactericidal properties as well as being an anti-inflammatory agent.[24]

As you see, these are extremely wide effects, and for this reason we should be sure that our skin is rich in ECGC at all times. Topical application enforces this wide collection of benefits to skin cells.

> "Unfermented tea leaves contain epi-gallocathechin gallate (EGCG), a potent singlet-oxygen antioxidant."

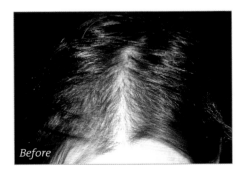

*This patient demonstrates the value of using topical vitamin A in low doses with higher doses of biotin for three months.*

## ROSEMARY EXTRACT (CARNOSIC ACID)

Rosemary is a popular herb in western diets, and in skin preparations it reduces UV damage to skin.[25] Rosemary extract combined with vitamin E and vitamin C becomes more powerful in suppression of collagenases.[26] Vitamin E also increases the stability of the rosemary extract carnosic acid and its uptake by cells. Carnosic acid is a good antioxidant and reduces erythema (redness) after UV exposure. It is also known to have anticarcinogenic properties.[27] Finally, rosemary extract is an antiseptic, thus it helps to preserve products from microbiological degradation.[28]

## BIOTIN

This is a water-soluble molecule that has been called vitamin B7 and also vitamin H. It is a cofactor in enzyme activity, meaning that it optimizes the actions of certain enzymes and is necessary for cell growth. Biotin strengthens hair, nails, and skin through its effects on metabolism.[29] Unfortunately, we still do not know enough about its full effects, but supplementation of the diet with biotin and its derivatives will help to optimise cellular health.[30] Biotin is generally found only in low doses in our food. Good sources of this vitamin however, are royal jelly, brewer's yeast, tomatoes, carrots, raspberries, strawberries, and walnuts.

## REFERENCE

1. Orallo, F., *Comparative studies of the antioxidant effects of cis- and trans-resveratrol.* Curr Med Chem, 2006. **13**(1): p. 87-98.
2. Potter, G.A., et al., *The cancer preventative agent resveratrol is converted to the anticancer agent piceatannol by the cytochrome P450 enzyme CYP1B1.* Br J Cancer, 2002. **86**(5): p. 774-8.
3. Mehta, R.G. and J.M. Pezzuto, *Discovery of cancer preventive agents from natural products: from plants to prevention.* Curr Oncol Rep, 2002. **4**(6): p. 478-86.
4. Khanna, S., et al., *Dermal wound healing properties of redox-active grape seed proanthocyanidins.* Free Radic Biol Med, 2002. **33**(8): p. 1089-96.
5. Newton, R.A., et al., *Post-transcriptional regulation of melanin biosynthetic enzymes by cAMP and resveratrol in human melanocytes.* J Invest Dermatol, 2007. **127**(9): p. 2216-27.
6. Lin, C.B., et al., *Modulation of microphthalmia-associated transcription factor gene expression alters skin pigmentation.* J Invest Dermatol, 2002. **119**(6): p. 1330-40.
7. Aziz, M.H., F. Afaq, and N. Ahmad, *Prevention of ultraviolet-B radiation damage by resveratrol in mouse skin is mediated via modulation in survivin.* Photochem Photobiol, 2005. **81**(1): p. 25-31.
8. Zhuang, H., et al., *Potential mechanism by which resveratrol, a red wine constituent, protects neurons.* Ann N Y Acad Sci, 2003. **993**: p. 276-86; discussion 287-8.
9. Binienda, Z.K., et al., *Assessment of 3-nitropropionic acid-evoked peripheral neuropathy in rats: neuroprotective effects of acetyl-l-carnitine and resveratrol.* Neurosci Lett, 2010. **480**(2): p. 117-21.
10. Juhasz, B., et al., *Resveratrol: a multifunctional cytoprotective molecule.* Curr Pharm Biotechnol, 2010. **11**(8): p. 810-8.
11. Dani, C., et al., *Antioxidant and antigenotoxic activities of purple grape juice--organic and conventional-in adult rats.* J Med Food, 2009. **12**(5): p. 1111-8.
12. Skaper, S.D., et al., *Quercetin protects cutaneous tissue-associated cell types including sensory neurons from oxidative stress induced by glutathione depletion: cooperative effects of ascorbic acid.* Free Radic Biol Med, 1997. **22**(4): p. 669-78.
13. Lee, E.J. and H.D. Jang, *Antioxidant activity and protective effect on DNA strand*

scission of Rooibos tea (Aspalathus linearis). Biofactors, 2004. **21**(1-4): p. 285-92.
14. Darvesh, A.S., et al., *Oxidative stress and Alzheimer's disease: dietary polyphenols as potential therapeutic agents.* Expert Rev Neurother, 2010. **10**(5): p. 729-45.
15. Kokotkiewicz, A. and M. Luczkiewicz, *Honeybush (Cyclopia sp.) – a rich source of compounds with high antimutagenic properties.* Fitoterapia, 2009. **80**(1): p. 3-11.
16. Marnewick, J.L., W.C. Gelderblom, and E. Joubert, *An investigation on the antimutagenic properties of South African herbal teas.* Mutat Res, 2000. **471**(1-2): p. 157-66.
17. McKay, D.L. and J.B. Blumberg, *A review of the bioactivity of South African herbal teas: rooibos (Aspalathus linearis) and honeybush (Cyclopia intermedia).* Phytother Res, 2007. **21**(1): p. 1-16.
18. Butt, M.S. and M.T. Sultan, *Green tea: nature's defense against malignancies.* Crit Rev Food Sci Nutr, 2009. **49**(5): p. 463-73.
19. Elmets, C.A., et al., *Cutaneous photoprotection from ultraviolet injury by green tea polyphenols.* J Am Acad Dermatol, 2001. **44**(3): p. 425-32.
20. Nihal, M., et al., *Anti-proliferative and proapoptotic effects of (-)-epigallocatechin-3-gallate on human melanoma: possible implications for the chemoprevention of melanoma.* Int J Cancer, 2005. **114**(4): p. 513-21.
21. Balasubramanian, S. and R.L. Eckert, *Keratinocyte proliferation, differentiation, and apoptosis--differential mechanisms of regulation by curcumin, EGCG and apigenin.* Toxicol Appl Pharmacol, 2007. **224**(3): p. 214-9.
22. Makimura, M., et al., *Inhibitory effect of tea catechins on collagenase activity.* J Periodontol, 1993. **64**(7): p. 630-6.
23. Vayalil, P.K., et al., *Green tea polyphenols prevent ultraviolet light-induced oxidative damage and matrix metalloproteinases expression in mouse skin.* J Invest Dermatol, 2004. **122**(6): p. 1480-7.
24. Toda, M., et al., *[Antibacterial and bactericidal activities of Japanese green tea].* Nippon Saikingaku Zasshi, 1989. **44**(4): p. 669-72.
25. Martin, R., et al., *Photoprotective effect of a water-soluble extract of Rosmarinus officinalis L. against UV-induced matrix metalloproteinase-1 in human dermal fibroblasts and reconstructed skin.* Eur J Dermatol, 2008. **18**(2): p. 128-35.
26. Offord, E.A., et al., *Photoprotective potential of lycopene, beta-carotene, vitamin E, vitamin C and carnosic acid in UVA-irradiated human skin fibroblasts.* Free Radic Biol Med, 2002. **32**(12): p. 1293-303.
27. Minnunni, M., et al., *Natural antioxidants as inhibitors of oxygen species induced mutagenicity.* Mutat Res, 1992. **269**(2): p. 193-200.
28. Aruoma, O.I., et al., *An evaluation of the antioxidant and antiviral action of extracts of rosemary and Provencal herbs.* Food Chem Toxicol, 1996. **34**(5): p. 449-56.
29. Cashman, M.W. and S.B. Sloan, *Nutrition and nail disease.* Clin Dermatol, 2010. **28**(4): p. 420-5.
30. Ames, B.N., *A role for supplements in optimizing health: the metabolic tune-up.* Arch Biochem Biophys, 2004. **423**(1): p. 227-34.

# Chapter 13
# PEPTIDES

Peptides are 'mini' proteins. They are a complex and exciting group of ingredients and are found today in many skin- and hair- care products on the shelves in stores. Over the past 15 to 20 years more and more peptides have become available. A peptide is a tiny building block for a larger protein and is composed of smaller building blocks called amino acids like glycine, cystine, or alanine. Each amino acid in turn consists of a nitrogen compound with an acid group, so they are made of nitrogen, hydrogen, carbon, and oxygen. There are only 20 amino acids that we use in our body, and from those amino acids we can create groups of amino acids in a chain called peptides. By using as few as only two of any of these 20 amino acids to make a di-peptide, we have the possibility of about 400 variants; the possible permutations of all 20 are enormous.

If we consider only the peptides that have fewer than ten amino acids, which we classify as oligopeptides (from Greek prefix *oligo-*, 'having few or little'), there are literally hundreds of thousands of possible peptides. Not all these oligopeptides will have a function, as they need to fit into receptors on cells and such receptors are much more limited in number.

Polypeptides (*poly-* from Ancient Greek *polus*, meaning 'many, much') are peptides containing between ten and 100 amino acids. The number of possible different combinations of amino acids is unimaginably huge. Again, not every possible polypeptide has a function in the body. Polypeptides in turn form even longer chains and link up to create proteins which constitute about 15 per cent of the human body and make up the largest 'solid' compartment. Water makes up roughly 70 per cent, leaving fats and carbohydrates together to amount to another 15 per cent in a *lean* body.

We find peptides naturally in all forms of life. Some peptides have been well known to us for a long time – insulin for instance, is a peptide and consists of two peptide chains with a total of 51 amino acids. Most of the messenger chemicals such as growth factors that are used to communicate between cells are peptides. We have known about peptides and proteins for many years, but scientists only started research in peptides about 50 years ago.

Peptides have a crucial importance in regulating the functions of cells. There are naturally active peptides such as the aforementioned insulin and many enzymes and coenzymes, cytokines and growth factors. The functional molecules of growth hormone are known as the 'somatomedins' and these are also peptides. We know that the body normally uses peptides all the time, but these natural peptides are unfortunately not generally available for topical application. Some peptides, like insulin, are so large they cannot penetrate the stratum corneum. Glutathione, one of the most important antioxidants inside skin cells, is a tri-peptide (from cysteine, glutamic acid, and glycine) and cannot exist outside the cell.

Pharmaceutical laboratories have engineered a variety of peptides that function for treating cancer, hormonal problems, and infections, illustrating just how wide and far this class of molecule is being pursued to provide effective treatment in many spheres of human health.

## NAMING PEPTIDES

When you look through the names of the peptides you will see that there is an initial designator telling you how many amino acids are in the peptide: *di* (two), *tri* (three), *tetra* (four), *penta* (five)-*peptide*. These conventional prefixes are all Greek numbers. The current nomenclature for peptides is based on the number of amino acids (not the different types) in the peptide, and then there is a number usually after the word *peptide* to distinguish it. This unfortunately creates difficulty and confusion in naming these molecules.

One can have peptides of different properties but the same number of amino acids, and consequently they have the same general chemical name, for example a pentapeptide-3 can describe both Leuphasil and Violox—two very different peptides when it comes to their actions on cells (see below). A more exact naming system will ultimately become necessary.

## PEPTIDES AND SKIN

When one closely examines the results with topical vitamins A and C, one

> "*Most of the messenger chemicals such as growth factors that are used to communicate between cells are peptides.*"

Before

After

*Results of peptides creams with peptides – UV simulated picture*

realizes that the changes could only come about with the assistance of growth factors or maybe the release of peptides to act as intercellular messengers. Peptides can stimulate the activity of growth factors in the right balance to stimulate fibroblasts to produce collagen, elastin, and GAGs. They also can stimulate keratinocyte growth patterns and thicken the epidermis. The synergy between vitamins A and C and appropriate peptides is probably the most powerful rejuvenating regime when combined with topical antioxidants. Enhancing penetration should produce even better results.

Peptides and growth factors have become one of the most exciting fields of skin-care research today, but the peptides we are interested in contain smaller and highly specific groups of amino acids that have special messenger functions between skin cells. Dipeptides are the smallest peptides and consist of two amino acids joined together. Peptides give us a chance to stop the natural degradation that occurs from photoageing as well as chronological ageing (chronoageing).

We should concentrate specifically on these peptides. Some peptides are similar to those normally found in the body and may therefore be classed as bioequivalent. Others are synthetic and their effects have been demonstrated by intensive research.

Peptides work by sending messages from one cell to another, for example from the keratinocyte to the fibroblast to make more collagen. Some peptides may be very large and cannot penetrate skin unless one employs penetration enhancement techniques such as cosmetic needling or low frequency sonophoresis with iontophoresis to help them through the barrier of a healthy stratum corneum layer. A further difficulty is that the skin has enzymes that may destroy peptides, hence many formulations have to protect peptides by chemical means.

The discerning public is beginning to demand more highly refined effective ingredients rather than the relatively crude plant derivatives that have been the mainstay of cosmetics in the past. Those natural extracts may contain many different molecules and even

*"Peptides work by sending messages from one cell to another, for example from the keratinocyte to the fibroblast to make more collagen."*

peptides, some of which might be active, but at concentrations that are far too low to work. The best these products can do is to be non-active or non-irritant. The addition of refined peptides to balanced and finely tuned topical preparations offers a way to make specific cell changes without increasing the possibility of skin reactions.

Peptides for cosmetics can generally be classed as those that are:

1. Signal peptides: messengers between cells, for example those peptides that stimulate the formation of collagen and elastin. Matrixyl and copper tripeptides are examples of these. Many of these effective peptides are similar to the terminal ends of normal collagen fibres or elastin fibres.[1]
2. Inhibitors of the enzymes that destroy collagen and elastin.[2, 3]
3. Neurotransmitter-affecting peptides that are active on nerves:[4] for example, peptides that inhibit the release of chemicals from the nerves to stimulate muscle fibres. A few examples are Argireline, Leuphasil, Violox, and Syn-Ake. These have a botulinum-like effect of reducing muscle tone.[5, 6]
4. Carrier peptides like the copper peptides that convey copper into the skin.[7]
5. Peptides that alter pigmentation by changing the production of melanin.[4]
6. Peptides with antioxidant activity.[8]

## THE ADVANTAGE OF USING PEPTIDES OVER WELL-KNOWN AND RESEARCHED GROWTH FACTORS

Many growth factors are peptides, and peptides in turn stimulate growth factors. We are in the relatively early investigative phase of understanding growth factors, but we have learned some interesting details about them. Very often they do not work in isolation and require some assistance from other growth factors. At this stage we believe that different ratios of growth factors can have very different results, and no one knows the ideal ratios. Furthermore, growth factors may promote melanoma formation[9] Growth factors are generally quite complex and usually moderately large molecules, typically proteins that do not easily cross the skin barrier, so the temptation is to enhance penetration with cosmetic skin needling, low-frequency sonophoresis, and/or iontophoresis. We cannot predict the degree of enhancement of growth factors and the ratios even when so-called ideal ratios[10] have been applied. At the present stage, the lack of clear information favours the use of simpler peptides that have been studied more intensively so that we know their ideal concentrations. The peptides are actually precursors of the growth factors that produce the results. They stimulate cells to produce ideal ratios of growth factors and hence are preferable at this stage to using growth factors by themselves. Its may also be safer to use peptides.

## WHAT ARE THE BENEFITS PEPTIDES OFFER (OR DON'T OFFER)?

Peptides offer a wide range of benefits if they are well formulated with penetrant enhancers and/or used with penetrant-enhancing techniques. The benefits will depend on the type of peptide and the concentration. If they are well formulated by knowledgeable chemists, the difference should become obvious to the user within several weeks. However, they cannot replace the primary ingredients required to normalize cell function in the form of vitamins A, C, and E. They will be less effective if used individually in preference to a well-balanced and highly-developed 'complexed' formulation.

## WHAT ARE THE LATEST DELIVERY SYSTEMS TO TAKE PEPTIDES TO THE PLACES WHERE THEY WORK?

Most of the peptides that we have heard about are relatively small. However, we have to bear in mind that something like Argireline has a molecular weight of 888 and, because of the acetyl component, can penetrate skin better than ascorbic acid. Ascorbic acid, with a molecular weight of only 176, finds it difficult to penetrate through skin because it is water-soluble.

Retinyl palmitate (the storage form of vitamin A) has a molecular weight of 524 and penetrates relatively well compared to retinoic acid, which has a molecular weight of 300. It is not necessarily just the size of the peptide that is important but also its solubility. One has to know the relative solubility of the formulation in lipids versus its solubility in water. Water-soluble properties make it difficult for peptides to penetrate skin. If, however, they are lipid-based peptides they will penetrate skin much more readily. Peptides which are simply too large will not get into skin without help. That is why we have to look at means and mechanisms to enhance the penetration of larger molecules through skin.

## CHEMICAL PENETRATION ENHANCEMENT

There are chemical techniques in the formulation which enhance penetration. Generally the average cosmetic cream, which has little or no activity on the cellular level, does not require enhancement of penetration because of the lack of any physiological activity. Cosmetics with cell-active molecules that make a difference to skin need careful formulation to ensure that the active chemicals actually get into the cells to make a difference.

Certain chemicals are used as penetrant enhancers, yet many people are not aware of the specialized actions of these compounds. Many people prefer to think that elegant constructs like liposomes are the only way to get more active molecules into the skin. Little attention is paid to rather 'ordinary' molecules like ethoxy-ethanol and the special emulsifiers, which are in fact much more important in this context. Paradoxically, there is a trend to

> "*Generally the average cosmetic cream, which has little or no activity on the cellular level, does not require enhancement of penetration because of the lack of any physiological activity.*"

> *"These microscopic holes allow larger molecules such as peptides to literally 'fall through' the skin barrier and immediately become available to living cells below."*

avoid emulsifiers and use formulations that specifically do not penetrate the stratum corneum. Those types of products are for surface treatments only, whereas when one needs to get molecules like peptides into the skin one has to use chemical penetrant enhancers. Peptide products should definitely be formulated to ensure that the peptide is able to penetrate skin and also be bioavailable. Peptides are often breakdown products of proteins; they have the potential to be broken down further, and consequently they have to be stabilized lest they deteriorate further in a topical cream. The normal chemical barrier of the skin may actually promote decomposition of peptides, and so before we can concentrate on the penetration of the peptides we have to ensure that they are intact and available for penetration.

## PHYSICAL PENETRATION ENHANCEMENT

Another simple method to enhance penetration is to bypass the stratum corneum, which is the obstruction to the penetration of molecules through skin. This can be easily done with a special device used since 2000 that creates temporary micropunctures through the stratum corneum.

These microscopic holes allow larger molecules such as peptides to literally 'fall through' the skin barrier and immediately become available to living cells below. They can additionally be transferred to the cells lying even deeper, which are the target cells for the peptides. Potentially this system can enhance penetration as much as 100 times better than simple topical application, but this does require an intensive treatment of the skin for about five to ten minutes.

Cosmetic Roll-CiT/ Focus-CiT are used immediately before products containing peptides such as Argireline, Leuphasil, Violox and Syn-ake to reduce muscular activity within a quarter to half the time required by simple topical application.

The important question is whether enhanced penetration of these peptides has been shown to be safe. There is only one company in the world with an experience going back to 2000. Clinical experience in over 20,000 people using cosmetic needling since 2000 and using the peptide-rich products produced by Environ Skin Care, has revealed consistently safe and healthier skin.[11] No untoward effects have been seen except for retinoid reactions, which are temporary changes and disappear as the skin becomes adapted to normal levels of vitamin A; they are not due to any side effects of peptides.

Another method to enhance penetration is iontophoresis, which is the use of a direct charged current (either negative or positive) to help propel the peptide into the skin. Amino acids carry either a neutral, positive, or negative charge and, as a result, peptide charge is dependent on the charge of the various amino acids in it. If we know the charge of the peptide then we can use an appropriately similar electrical charge

Before

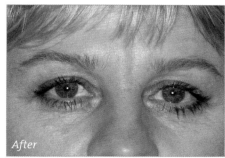
After

*It's difficult to show reduction in frowning with a simple photograph but this case shows the patient trying to frown on both occasions. She has achieved this by using a cocktail of three neuropeptides argireline, leuphasil and violox after cosmetic needling of her forehead.*

(iontophoresis) to propel the peptide into the skin. By doing this we can enhance penetration by up to four times more than simply applying the product onto the skin,[12] even for treatment periods of only ten minutes.[13] An important advantage of this technique is that we can get even larger molecules – even insulin -through skin.[14] Insulin is large and has a molecular weight of 5808 Dalton but with iontophoresis one can get this large peptide into the bloodstream and even control the blood sugar levels in diabetics.

Low-frequency sonophoresis (LFS) is probably the most exciting way to enhance penetration. Introduced by Mitragotri and associates in 1996 when he proposed that low-frequency 'ultrasound' (not the conventional ultrasound) could enhance penetration by 40 times compared to simple application to the skin.[15]

The combination of iontophoresis and LFS is the most effective method ever described to enhance the penetration of active peptides that relax the frown and smile muscles, and many patients notice a change within hours of having the treatment. Ideally these peptides can be administered by a combination of iontophoresis and low-frequency sonophoresis with a home-based device either daily or two to three times a week.

Alternatively, one can achieve and maintain results with topical application using cosmetic-needling of the horny layer of the skin.

One can feel confident in using these three techniques with the particular peptides because the products have been formulated and studied for almost 20 years. They have been found to be the most effective, consistent, and predictable methods to enhance penetration of molecules and rejuvenate skin.[11, 16-20]

## THE PEPTIDE DILEMMA: WHICH IS THE BEST PEPTIDE TO USE?

Since peptides are being specially designed and introduced all the time, it is quite difficult to be sure what the latest and best peptides are. It is preferable to comment only on some that have been the subject of medical research over and above the manufacturer's in-house research, which may potentially be biased. However, in an extremely competitive world with restricted cosmetic claims, manufacturers are doing intensive research to validate new ingredients. Here are a few well-known peptides.

## COPPER TRIPEPTIDES

Copper tripeptides have been used for more than 15 years because of their observed effects on wound healing. This effect has been extrapolated to skin rejuvenation. Part of their action comes from the inhibition of enzymes like collagenases and other matrix-metalloproteinases.[7]

## MATRIXYL (PALMITOYL PENTAPEPTIDE)

Matrixyl (palmitoyl pentapeptide) was introduced in 1999 and gained popularity when it was shown in 2000 to give effects comparable to retinoic acid.[21] Since vitamin A, as retinoic acid, is the only molecule that has been proven to give a broad-scale real rejuvenation of the skin, this was an important comparison. While the changes in collagen and elastin production were similar to that caused by vitamin A, there was no demonstrated change to the photodamaged DNA, and that is a very important difference. However, this indicates the potential synergy between vitamin A and Matrixyl. At this stage Matrixyl continues to be one of the important peptides for rejuvenating skin and has been subsequently modified e.g. as Matrixyl 3000 which is even more effective.[22]

## ARGIRELINE (ACETYL HEXAPEPTIDE-3)

Initially people thought it was impossible that a peptide could be absorbed in such a low concentration through the skin and still be effective on muscles relatively so far away. However, these peptides can work at extremely low concentrations in the order of one part per billion.

Argireline (acetyl hexapeptide-3) acts on neuromuscular junctions and reduces the contraction of muscles. It does this through its action on the 'SNAP' complex and through preventing acetylcholine vesicles from being released through the nerve membrane in much the same way as botulinum toxin (Botox) injections. Acetylcholine is the molecule that stimulates a muscle to contract. Because acetylcholine cannot be released into the space between the nerve ending and the muscle receptor plates, muscles cannot contract. Argireline is an ideal companion for Matrixyl because it reduces the action of muscles that help to make wrinkles, while Matrixyl increases the collagen in the skin against the background of a more relaxed skin. The new collagen is therefore laid down in the relaxed stress pattern, rather than the contracted stress pattern. This will logically have the effect of a smoother skin.

## LEUPHASIL (PENTA-PEPTIDE-3)

Leuphasil (pentapeptide-3) works by blocking the special receptor for calcium that is normally situated on the cell walls of nerves. This reduces

> "The different molecules working synergistically are more powerful than the peptides on their own."

the inflow of calcium to the nerve, and as a result there are lower levels of intracellular calcium, which prevents the nerve membrane from opening the acetylcholine vesicles. It is easy to see that Leuphasil augments the effects of Argireline in reducing the activity of the muscles close to the skin.

## VIOLOX (ALSO PENTAPEPTIDE-3)

Violox (also pentapeptide-3) blocks the contraction of muscles by blocking the receptors for acetylcholine on the surface of the muscles. Violox prevents the flow of sodium into the muscle cells through the special sodium channels, which are stimulated by acetylcholine and cause muscle contraction. (It is much the same as the poisonous dart made of curare used by Amazonian Indians to immobilize their prey.) Violox is an extremely expensive peptide.

Great success has been achieved with a combination of (Argireline acetyl hexapeptide-3.) and Leuphasil (pentapeptide 3) and Violox (pentapeptide-3) or SNY-ake (dipeptide diaminobutyroyl). The different molecules working synergistically are more powerful than the peptides on their own.

## PALMITOYL OLIGOPEPTIDE

Palmitoyl oligopeptide is a tripeptide that stimulates the production of collagen IV and other fibres in the basement membrane between the epidermis and dermis. The basement membrane is a critically important structure, and, because sunlight and other forces damage it as we get older, it loses the

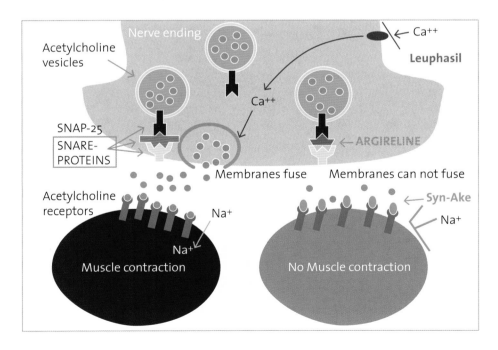

*Diagram of action of Argireline combined with Leuphasil and Syn-Ake*

ability to anchor the epidermis to the dermis effectively. The regulating role the basement membrane has in maintaining the dermal papillae to keep the epidermis well anchored on to the dermis, thereby increasing the interface between these two layers, is gradually lost. An important consequence of these undulations, called rete pegs and rete ridges, is that the epidermis is well nourished even though it does not have its own blood supply. As the basement membrane weakens, the interface gets smaller and the epidermis becomes malnourished and thinner. Another advantage of palmitoyl oligopeptide is that it also stimulates the formation of elastin, which lies just below the epidermis.

### OTHER PEPTIDES

There are many other peptides and most certainly there are important ones that have not been discussed. Information on peptides could fill several volumes on its own. We need to understand only general principles in this very complex field.

### WHAT ARE THE LATEST ADVANCES AND FUTURE APPLICATIONS OF PEPTIDES?

It is difficult to be certain exactly which of the newly introduced peptides are going to be the most successful, but one thing is certain: peptides by themselves may prove to help make significant changes in skin, yet when they are combined with vitamin A (retinoids), vitamins C and E and other ingredients, I believe much more impressive results will be achieved. Some peptides may extend the life of cells; others have been recommended for cellulite, for hair growth, and for control of pigmentation. We will see greater and greater value in newly developed peptides as scientists learn to design more active peptides that 'speak the language' of the cells.

A search of the medical literature reveals that there is a remarkable lack of clinical studies on topical peptides. Most of the studies have been reported in the cosmetic chemistry literature and from the companies making the ingredients, which unfortunately attaches an automatic bias to these papers.

### REFERENCE

1. Bauza, E., et al., *Collagen-like peptide exhibits a remarkable antiwrinkle effect on the skin when topically applied: in vivo study*. Int J Tissue React, 2004. **26**(3-4): p. 105-11.
2. Brassart, B., et al., *Conformational dependence of collagenase (matrix metalloproteinase-1) up-regulation by elastin peptides in cultured fibroblasts*. J Biol Chem, 2001. **276**(7): p. 5222-7.
3. Antonicelli, F., et al., *Elastin-elastases and inflamm-ageing*. Curr Top Dev Biol, 2007. **79**: p. 99-155.
4. Lupo, M.P. and A.L. Cole, *Cosmeceutical peptides*. Dermatol Ther, 2007. **20**(5): p. 343-9.
5. Wang, Y., et al., *The anti-wrinkle efficacy of argireline, a synthetic hexapeptide, in Chinese subjects: a randomized, placebo-controlled study*. American journal of clinical dermatology, 2013. **14**(2): p. 147-53.
6. Wang, Y., et al., *The anti-wrinkle efficacy of Argireline*. Journal of cosmetic and laser therapy : official publication of the European Society for Laser Dermatology, 2013. **15**(4): p. 237-41.
7. Mazurowska, L. and M. Mojski, *Biological activities of selected peptides: skin penetration ability of copper complexes with peptides*. J Cosmet Sci, 2008. **59**(1): p. 59-69.
8. Mendis, E., N. Rajapakse, and S.K. Kim, *Antioxidant properties of a radical-scavenging peptide purified from*

8. *enzymatically prepared fish skin gelatin hydrolysate.* Journal of agricultural and food chemistry, 2005. **53**(3): p. 581-7.
9. Berking, C., et al., *Basic fibroblast growth factor and ultraviolet B transform melanocytes in human skin.* Am J Pathol, 2001. **158**(3): p. 943-53.
10. Sundaram, H., et al., *Topically applied physiologically balanced growth factors: a new paradigm of skin rejuvenation.* Journal of drugs in dermatology : JDD, 2009. **8**(5 Suppl Skin Rejuenation): p. 4-13.
11. Fernandes, D.B., *Making sense of the cosmeceutical concept* 2013(chapter 19): p. 228-238.
12. Sloan, J.B. and K. Soltani, *Iontophoresis in dermatology. A review.* J Am Acad Dermatol, 1986. **15**(4 Pt 1): p. 671-84.
13. Akomeah, F.K., G.P. Martin, and M.B. Brown, *Short-term iontophoretic and post-iontophoretic transport of model penetrants across excised human epidermis.* Int J Pharm, 2009. **367**(1-2): p. 162-8.
14. Batheja, P., R. Thakur, and B. Michniak, *Transdermal iontophoresis.* Expert Opin Drug Deliv, 2006. **3**(1): p. 127-38.
15. Mitragotri, S., D. Blankschtein, and R. Langer, *Transdermal drug delivery using low-frequency sonophoresis.* Pharm Res, 1996. **13**(3): p. 411-20.
16. Fernandes , D.B., *Topical delivery of cosmetic ingredients by Sonophoresis and Iontophoresis.* Science and Applications of Skin Delivery Systems, 2008. **edited by J. Wiechers**(chapter 20).
17. Fernandes, D.B., *Evolution of Copsmeceuticals and Their Application to Skin Disorders, Including Ageing and Blemishes.* Dermatological and Cosmeceutical Development: Absorption Efficacy and Toxicity, 2007. **Edited by Kenneth A Walters, Michael S. Roberts**(Chapter 4): p. 45-60.
18. Fernandes , D.B., *Pre and post-operative skin care.* Aesthetic Surgery of the Facial Mosaic, 2006. **edited by D.E. Panfilov**(chapter 62): p. 492-502.
19. Aust, M.C., et al., *Percutaneous collagen induction therapy: an alternative treatment for scars, wrinkles, and skin laxity.* Plast Reconstr Surg, 2008. **121**(4): p. 1421-9.
20. Fernandes , D.B., *The Holistic Scarless Rejuvenation of the Face.* Miniinvasive Face and Body Lifts – Closed Suture Lifts or Barbed Thread Lifts, 2013. **edited by Nikolay Serdev**(Chapter 9): p. 195 – 212.
21. Robinson, L.R., et al., *Topical palmitoyl pentapeptide provides improvement in photoaged human facial skin.* Int J Cosmet Sci, 2005. **27**(3): p. 155-60.
22. Chirita, R.I., et al., *Development of a LC-MS/MS method to monitor palmitoyl peptides content in anti-wrinkle cosmetics.* Analytica chimica acta, 2009. **641**(1-2): p. 95-100.

# Chapter 14
# OTHER MOLECULES TO REJUVENATE SKIN: ALPHA- AND BETA-HYDROXY ACIDS AND OTHERS

Alpha-hydroxy acids (AHA) have been the most popular ingredients of so called 'cosmeceuticals' for many years, but now they are losing their lustre. By themselves their physiological activity poses problems, but they do have specific benefits and are ideally combined with topical vitamin A. The same is true for beta-hydroxy acids (BHA), but they each have other specific effects that are good for skin. The AHAs are active cosmetic ingredients that are not generally intended for younger people with healthy skin, and should be used only from about the age of 25 and older. When young people have acne or some other problem of excessive keratinization then AHAs and BHAs can be very effective.

## THE CHEMISTRY OF HYDROXY ACIDS

Hydroxy acids are natural organic acids and are grouped as alpha or beta hydroxy acids because of a slight difference in their molecular structure. AHAs are also known as fruit acids (see figure 5.1), but many of them, like glycolic and lactic acid, are not derived from fruit. Glycolic acid actually comes from sugar cane and hence is not a fruit acid in the true sense. Lactic acid comes from milk but is also found in tomatoes, whilst maleic, pyruvic, and citric acids are derived from fruits.

Glycolic and lactic acid are the most commonly used ingredients in cosmetics, but many cosmetic houses claim to have 'superior' acids found in various fruits

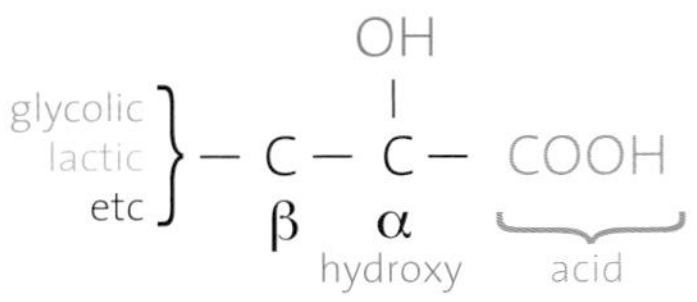

This illustration gives the basic chemical formula for alpha (a) hydroxy acids and shows the relative position of the acid group COOH) to the hydroxy (OH). The positioning of the OH "softens" the acidic effects. If the OH is on the b position then the acid becomes "stronger" as in beta hydroxy acids.

like strawberries. Another marketing claim that should be clearly understood is that glycolic acid, used in all cosmetics (even so-called natural products), is not natural but is synthesized in a laboratory. Lactic acid, on the other hand, may be either natural or 'nature-identical but synthesized'. Natural L-lactic acid has different properties from the synthetic DL-lactic acid (see explanation below). We should only use nature-identical synthetic L-lactic acid.

The molecular structures of these various acids may be very different. Some are very small like glycolic acid, and it is claimed that glycolic acid penetrates deeper and faster, while others claim that acids with a larger structure like lactic acid or pyruvic acid have a safer and more superficial action. However, there is no clear-cut scientific evidence to substantiate these claims. The size of the molecule is not as important as other chemical properties, but that involves complicated chemistry outside the scope of this text.

## THE CHEMISTRY OF AHAS AND BHAS

The name alpha-hydroxy and beta-hydroxy acid comes from a shorthand designation to indicate where the hydroxyl (OH) group is situated on the molecule. These acids are called carboxylic acids and there are very many different carboxylic acids. (See diagram).

All of the hydroxy carboxylic acids (-COOH) have an attached hydroxyl group (OH) added as in the diagram. When the OH group is situated on the alpha carbon, the acid effect is weaker than when the OH is situated on the beta position. The presence of this hydroxy group 'softens' the effects of the acid that would otherwise be too strong or even poisonous. The softening effects are mediated through complex electrical events involving electrons that defuse the power of the acid radical. One or two atoms in specific positions change the properties of these molecules significantly.

Beta-hydroxy acids are stronger acids because there is less 'softening' of the acid component of the molecule by the hydroxyl group (-OH) on a more distant carbon atom. The arrangement of the hydroxyl radical to the acid component is fixed in all the alpha- or beta-hydroxy acids, while the other part of the molecule may be quite

*"However, paradoxically, prolonged use of AHAs may aggravate dry skin because of damage to the horny layer, creating exactly the opposite condition."*

variable. That is why some of the acids are very short and have a low molecular weight, whereas others are much longer and have a high molecular weight. All of this is chemically very technical, but the important thing is that to belong to the same chemical group certain specific properties have to be present, while other parts may differ a great deal without altering the basic behaviour of the molecules. There are many different forms of AHA or BHA molecules, like a very large extended human family where the members share a common name and cultural heritage, yet have different personalities.

## THE HISTORY OF ALPHA-HYDROXY ACIDS

One may be forgiven for thinking that AHAs are very modern, but it seems the ancients knew about their beneficial effects. Cleopatra bathed in asses' milk (lactic acid), while Roman women used wine (acetic acid) to wash their faces and promote a healthy complexion. There are also rumours about Madame Pompadour, the mistress of Louis XV of France, who was said to bathe in champagne, which also has AHA properties for improving the skin. Similarly, poultices of fruit or vegetables like cucumber and strawberries may have had similar properties when applied to skin.

The modern scientific basis of AHAs started in the 1960s when Dr Van Scott in Philadelphia noticed that dry skin conditions responded very well to glycolic acid. Slowly his influence spread, and by the end of the 1980s it had become apparent that this acid could have a very positive effect on the skin.

There is a lot of wild publicity about the use of AHAs. Some people have been duped and deluded into believing that AHAs make radical changes to the skin on their own, and worse, can be used year after year with no ill effects. There were also interesting claims that it would make plastic surgery unnecessary. These are simply false promises created by marketing departments. Some people say that to be highly effective the AHAs have to be at a certain pH and of a certain minimum strength. Again, this has to be understood from a chemical point of view. The lower the pH (the higher the concentration of hydrogen atoms), the more effective the AHA is, but it is also more corrosive. At approximately pH 3.5 one has adequate AHA activity and a good level of safety. The higher the pH is, the safer the product becomes, but at pH over six there is unfortunately almost no acid effect on skin.

## HOW ALPHA-HYDROXY ACIDS WORK

The most obvious effect is that alpha-hydroxy acids promote natural desquamation (exfoliation of the dead and dying cells) of the horny layer of the skin. They dissolve the alkaline desmosomes bonds (that is, 'the 'cement' or glue) between these cells.[1] This allows the cells to flake off evenly and at a faster rate. The cells beneath are 'newer' (but they are still dead cells), fresher, and more hydrated. Their higher water content gives a softer and smoother appearance to the skin.[2] AHAs also compact the horny layer. Desquamation may continue for up to two weeks after stopping the use of AHA products.

AHAs, like glycolic and lactic acid, promote growth of the basal-layer cells.[3] They seem to promote healthier cells, and part of this action may be because unhealthy cells are less acid resistant and are selectively removed, while the healthier cells resist the acid and multiply and take over the place of the unhealthy cells. This is a sort of selective removal of weaker cells. Healthier skin becomes more waterproof, and so dry skin conditions are relieved. However, paradoxically, prolonged use of AHAs may aggravate dry skin because of damage to the horny layer, creating exactly the opposite condition.

In many cases, irregular pigmentation of the skin can be diminished,[4] but this effect may not be as strong as it would be with vitamin A. The reason for this is also different. Vitamin A affects the manufacture of melanin, whereas the AHAs act only by chemical action on melanin in the keratinocytes. The melanin bleaches from dark brown or black to a lighter colour because of the low pH. Lactic acid seems to have a specific lightening effect on pigmentation blemishes by its action on tyrosinase.[5]

The AHAs penetrate to some degree into the sebaceous follicle, where the build-up of dead cells is associated with acne conditions. The AHAs remove dead cells, thereby unblocking the follicle and preventing a build-up.[6] The initial improved moisturization extends down into the dermis and the skin gets better turgor, so wrinkles are less obvious. AHA and specifically lactic acid[7] promote formation of glycosaminoglycans in the dermis,[6] which adds to the water retention properties of the skin. Fibroblast cells produce more hydroxyproline, an essential amino acid that is a precursor to collagen, and more collagen is laid down in the process.[8] Because of the physical effect of stinging the skin, AHAs also improve the blood flow, by a widening of vessels through reflex nerve activity. Some carboxylic acids act as strong antioxidants, for example polyhydroxy acids and ascorbic acid (vitamin C)[6].

Many of these effects are very similar to vitamin A, and some people believed that AHAs were an alternative to vitamin A. Nothing can be further from the truth, even though, ironically enough, vitamin A acid is a carboxylic acid! Vitamin A works on different receptors and has its main effects on the DNA, whereas AHAs work at other sites, but certainly not on the DNA. The fact that they work in two different places is, however, an excellent reason for combining them to get the best effects.

In the early days of 'cosmeceutics' there were some irresponsible claims that AHAs, in contrast to vitamin A,

> "Many of these effects are very similar to vitamin A, and some people believed that AHAs were an alternative to vitamin A. Nothing can be further from the truth."

could be safely used while tanning! The fact is that AHA-treated skin is light-sensitive and must be protected from the sun. Skin with pigmentation problems should especially be protected against UVA when using AHA; that is why AHA should be used at night. AHAs should be seen as a means to an end, but not used daily year in year out forever. The most effective use is just occasionally as a 'booster'. There are some individuals who are able to use AHAs for prolonged periods without any detriment, but the majority probably benefit most from using them in 'treatment cures'. Use of AHAs periodically is best, twice or three times per week. There is no doubt that AHAs are valuable in good skin care, but they need to be used sensibly.

## A WORD ABOUT LACTIC ACID

Lactic acid is very likely the most important AHA that we have in the cosmetic arsenal. In contrast to glycolic acid, lactic acid is a very important part

$$\begin{array}{c|c} \text{COOH} & \text{COOH} \\ | & | \\ \text{OH} - \text{C} - \text{H} & \text{H} - \text{C} - \text{OH} \\ | & | \\ \text{CH}_3 & \text{CH}_3 \end{array}$$

*The two isomers of lactic acid to show their mirror image*

of our normal physiology – our skin, hair, blood, muscles, and other organs. In fact, we manufacture about 120 grams of lactic acid every day! [9] Lactic acid is found in two forms: (1) L-lactic acid, which is the natural human form of lactic acid and (2) D-lactic acid that is a mirror image of the natural lactic acid. Synthetic lactic acid consists of a mixture of L- and D-lactic acid. Synthetic (but nature-identical) lactic acid is synthesised and then the 'natural' L-lactic acid is separated from the D-lactic acid so that only natural L-lactic acid remains.

Natural lactic acid is metabolized faster by cells, and when used on the skin it causes less irritation but softens hard hyperkeratotic skin (skin with too much rough keratin protein) better and faster. When one examines the effects of lactic acid on ceramide (natural skin oil molecules involved in the skin barrier) production, D-lactic acid has minimal effects, but natural L-lactic acid has very impressive effects at the same dose. This is important because the increase in ceramides helps to make the skin more flexible and improves the stratum corneum barrier function.[2] Moreover, the increase in ceramides is important in treating dry skin in winter when the natural level of ceramides drop.[7]

Natural bio-identical lactic acid stings less than synthetic D-lactic acid, and significantly less than glycolic acid. Why this is so is not understood. Natural lactic acid has an important function in reducing pigmentation by suppressing the formation of tyrosinase. (Tyrosinase is the important enzyme involved in pigment production.)[5] It works well with other tyrosinase inhibitors like vitamin C. It is also antimicrobial and as such is a natural preservative.

(A) It is a powerful humectant.
(B) Lactic acid is an important and major part of the natural moisturising factors.
(C) It produces all the typical advantages of AHAs.
(D) It assists in maintaining the natural acid mantle.
(E) Lactic acid is an important skin-lightening ingredient.
(F) It is even safe to eat and has no negative impact on the environment.

In summary, lactic acid is a natural component of skin and has special properties:

Now for the big surprise! Natural lactic acid is derived from fermented sugar and not from milk.

## IS AHA AN ALTERNATIVE TO VITAMIN A?

The AHAs gained popularity just after retinoic acid was much discussed in the mid 1980s, and many detractors of retinoic acid hailed the AHAs as the successor to retinoic acid and claimed that they were much safer and more effective. They short-sightedly threw vitamin A out of the window, but one important fact remains true: no matter what the fashion may be, vitamin A is the *essential* vitamin for healthy skin.

It is strange that even though AHAs make the stratum corneum thinner because of the stripping of the outer layers of the horny layer, no one accuses them of 'thinning the skin'. On the other hand, critics say vitamin A makes the skin thinner, even though solid undeniable scientific evidence proves that the skin is even thicker with vitamin A than with the AHAs. The prejudice against vitamin A is hard to fathom. While it is true that not everyone needs such an active molecule as retinoic acid topically, we all need healthy doses of vitamin A in our skin to replenish the vitamin A deficiency that follows exposure to light.

On the other hand, prejudice against AHAs also exists. Some champions of retinoic acid claim that the AHAs have no great place in cosmetics, but both sides are also seriously mistaken. Clinical experience over many years has shown that the combination of vitamin A in its less aggressive forms and potent antioxidants together with AHAs, probably form the most important basic ingredients in skin care and are a formidable way to treat ageing skin. These three important ingredients, however, cannot be presented in the same cream without sacrificing the activity of one or the other, simply because vitamin A is destroyed at the pH that is required for AHA activity. The AHAs do not work at the pH that safeguards the vitamin A. The AHAs have a very low pH and work optimally at or below pH 4.5. At pH 6 the AHAs have negligible effects, but vitamin A is destabilised at below about pH 5.5. Paradoxically, one can safely mix the two creams in one's hand and make a mixture that will be effective, because the mixture must be used immediately so that there is no time for the above-mentioned interactions to occur.

*"No matter what the fashion may be, vitamin A is the essential vitamin for healthy skin."*

> "AHAs are particularly useful in acne treatment, because they increase the acidity of the skin and reduce the build-up of the horny layer clogging the ducts of the sebaceous glands."

## THE SIDE EFFECTS OF ALPHA-HYDROXY ACIDS

AHAs fortunately have few side effects except the following:

(A) Stinging on application.
(B) Transient redness in the early stages or if the preparation is applied too thickly.
(C) Flaking of the skin, usually only in the early stages.
(D) Some people may develop spots within the first one or two weeks.

These effects should not be interpreted as signs that someone is too sensitive for AHAs. However, problems will disappear once the skin adjusts. Prolonged use of AHAs without vitamin A support will eventually cause dry skin problems. This is the result of chronic stripping of the stratum corneum. This effect is similar to chronic mechanical abrasion of the surface of the skin.

## SENSITIVE SKIN AND ALPHA-HYDROXY ACIDS

AHAs are usually well tolerated by people with sensitive skins and people who seem to react to other creams. AHAs are particularly useful in acne treatment, because they increase the acidity of the skin and reduce the build-up of the horny layer clogging the ducts of the sebaceous glands. The acids are able to loosen and assist in removing sebaceous plugs known as whiteheads and blackheads. AHAs can be safely used by people with couperose (broken veins) and in those with rosacea.

## ALPHA HYDROXY ACIDS: BEST EFFECTS IN CONJUNCTION WITH VITAMIN A

As mentioned before, the use of AHAs with vitamin A should be standard practice. It is best not to use them on their own. Clinical trials have shown that improvement of the skin is more reliable and faster when the essential vitamins are used in conjunction with AHA, and other studies have confirmed that the skin is thicker when combination therapy is used.[10]

There is a belief that the AHAs facilitate the absorption of the vitamins by dissolving the lipid bilayers between the horny layer cells. This lipid-based 'glue' is made up of the proteins that were originally within the epidermal cells and were altered when the cells died. Before these materials are finally extruded from the cells they form the 'granules' seen in the granular layer that are extruded to form the lipid bilayers. The lipid bilayers act as 'glue' that holds the cells together and act as an important barrier to the loss of water from the skin and to various substances applied to the skin. By dissolving the 'glue' the AHAs allow vitamins to be easily transported through the horny layer into the stratum spinosum. Once in the spiny layer, the vitamins can easily get to the depths of the skin. The level of AHAs required to give better results varies. Some people need only a low level of AHA to get effects, whereas others need higher levels of AHAs.

An important and interesting phenomenon found in clinical trials was that people who have a pronounced sensitivity (retinoid) reaction to vitamin A can be rendered more tolerant to vitamin A by pre-treating the skin with AHAs. The reason for this is not quite clear yet.

## HOW TO USE ALPHA-HYDROXY ACIDS

People's skins are all different. Some skins will respond to rather low doses, whereas others require much higher concentrations. Very broadly speaking, AHA effects are usually created at concentrations above 2 to 3 per cent.

Use AHAs with the following general guidelines:

1. Transient stinging almost always occurs when they are applied. This is normal and is not a reason for concern.
2. Start on a low level to get effects without irritation of the skin. After an interval of weeks or months, the skin will tolerate a higher dose.
3. Always combine AHA with vitamin A products and antioxidants. The AHAs also have a place in body treatment where one can achieve a silky surface in people with rough skin. Lactic acid is most effective in treating rough and dry skin conditions. Neutralized lactic acid (like ammonium lactate) seems to be particularly active in softening callused skin.[11]

These modern chemicals combined in the correct way are effective and give us results previously never dreamed of. They may even help our present crop of young people to grow into their 50s looking as though they are still in their 30s. In time one hopes that we will get even better, less irritant, chemicals that may actually reverse the ravages of time. AHAs may be replaced by other chemicals or methods that work better, but the important point to bear in mind is that no matter what new chemicals may come into fashion, vitamins A and C and the antioxidants will always be essential in the true meaning of the word. There are no alternatives to these primary molecules, as the genetically inherited receptor systems on human cells need these precise molecules and cannot function normally with any others.

## IMPORTANT REMINDER

AHAs are really useful when used in moderation and as 'treatment spells'. They should not be used for an extended length of time. It is best not to torture skin into looking smooth, but rather to gently and physiologically encourage skin to be healthy. Everyone, especially people with darker pigmented skins or those with abnormal pigmentation, are advised to avoid the sun when using AHA preparations.

## BETA-HYDROXY ACIDS

Beta-hydroxy acids (BHA), like alpha-hydroxy acids, are derived from plants, fruit, and gluconamide, a sugar derivative. BHAs similarly break up the bonds (desmosomes) between the dead cells of the horny layer,[12] facilitating exfoliation and allowing new skin cells to emerge. Another important effect is that the lipid bilayers become more permeable and agents can penetrate more readily through this lipid barrier. BHA effects are very similar to AHA effects.

Citric acid and BHAs in the form of salicylic acid have been used either orally or topically for centuries to treat a variety of human diseases. Salicylic acid (chemically closely related to acetylsalicylic acid, commonly known as aspirin) is a widely researched substance that is becoming more and more important for preventing a range of illnesses in medicine, besides being an effective painkiller and anti-inflammatory medication. In skin care, synthetic salicylic acid has been studied for a long time and it is well known for its keratolytic action. Salicylic acid removes excess horny layer very efficiently and can penetrate down into the depths of the epidermis. In fact, salicylic acid can reach right into the dermis and can be absorbed into the bloodstream.[13]

In the past, the main reasons for application were to treat acne,[14] solar keratoses and thickened rough skin.[10] Many clinicians noted that skin appeared fresher after applying salicylic acid to the face and neck, and so salicylic acid was recognised for reversing ageing in skin.

The skin became thicker with increased amounts of glycosaminoglycans (the protein and sugar molecules that form the jelly-like substance between cells). Similar changes were observed with citric acid and beta-lipohydroxy acid.[15] The changes are similar to using retinoic acid, ammonium lactate, and glycolic acid, but histologically there are important differences from glycolic acid. The cells appear much more like younger cells after treatment with BHA than with glycolic acid.[16] More recently, research has shown that salicylic acid penetrates to lower dermal layers than glycolic acid but does not cause harm to underlying tissues or other body organs.[17]

The formulation is very important to determine the degree of penetration. In an optimal formulation (usually with an alcohol), salicylic acid can penetrate about 3 to 4 mm into the skin.[18] Acne and aged skin are not harmed by repeated applications of salicylic acid in low doses. At about 2 per cent salicylic acid there is no danger of adverse systemic effects; however, at higher doses there is a risk of salicylic acid toxicity.

Citric acid is an unusual molecule in that it includes both an AHA component and a BHA ending and possesses properties of both AHA and BHA. In clinical trials it has been shown to be an effective keratolytic, inducing a smoother stratum corneum. The epidermis becomes thicker and more viable with increased collagen and GAGs in the dermis.

## BHA FOR PHOTOAGEING

Professor Albert Kligman believed that BHAs are much better than AHAs. They do things that alpha-hydroxy acids cannot. One of those things is that a BHA like salicylic acid can help to treat pathological sugar deposits (glycation) in cells, which are a feature of ageing. That is why, for those who can, it is also a good idea to take a half tablet of aspirin every day! The research on the effects of salicylic acid and variants of BHAs indicate that they are effective in reversing premature ageing.

The exfoliation properties of salicylic acid that result in increased cell turnover and new cell production are not the only reasons for its ability to reverse premature ageing. It has been suggested that physiological changes occur and, among other things, tumour necrosis factor $\alpha$ (TNF-$\alpha$) is promoted, and that then mediates the changes in the skin.[15] TNF-$\alpha$ is a pro-inflammatory cell signal commonly involved in a large number of inflammatory reactions in the body.

Increased collagen synthesis, as has been shown with alpha-hydroxy acids, also occurs with beta-hydroxy acids. As a result, thin lines and wrinkles are reduced because underlying new cells and new collagen production in the dermal layers expand skin tissues and result in firmer, smoother-looking skin.

## BETA-HYDROXY ACIDS FOR ACNE

It has been known for years that salicylic acid penetrates through sebum, the substance produced by sebaceous glands. People with oily skins have more active sebaceous glands, and the excess sebum becomes mixed with the horny layer lining the follicles or sebaceous gland 'pores' and then hardens. A type of sebum cement is formed which obstructs the follicle and leads to whiteheads and blackheads and ultimately to acne lesions.

Salicylic acid and other BHAs can reverse this process by dissolving these keratin plugs.[14] Salicylic acid has the unique property of penetrating though pores, eliminating acne-causing bacteria, and helping to restore the irritated skin surrounding the pore. Salicylic acid lowers the pH and inhibits the growth of bacteria on the surface of the skin as well as in the follicles.

## SIDE EFFECTS OF BETA-HYDROXY ACIDS

The most common BHA is salicylic acid. At high doses (20 per cent and higher), salicylism (toxicity by salicylic acid) with ringing in the ears, nausea, and vomiting may occur, because salicylic acid penetrates the skin so effectively. In a study of side effects caused by hydroxy acids, salicylic acid caused virtually no irritation compared to glycolic acid, which can often cause stinging and burning sensations. Salicylic acid has superior exfoliating benefits compared to glycolic acid, and it is gentler to the skin. Long term use of salicylic acid does not eventually lead to dehydrated skin as often happens with unrelenting use of the AHAs. As with AHAs, one should avoid sun exposure when using a BHA. It is critical to protect the skin from UVA when using a BHA.

> "Citric acid and BHAs in the form of salicylic acid have been used either orally or topically for centuries to treat a variety of human diseases."

## A COMPARISON OF AHAS AND BHAS

1. AHAs are water-soluble and penetrate only the outermost epidermal skin layers. In contrast, BHAs are lipid-soluble and penetrate fairly rapidly to the thicker dermal skin layers located even 3 to 4 mm below the epidermis. That means that salicylic acid has the ability to penetrate to underlying tissue and have an influence on numerous cell layers, but otherwise does not cause harm.

2. Both beta- and alpha-hydroxy acids are effective in exfoliating skin, which causes increased cell turnover and the production of new skin cells.[19] Research indicates that beta-hydroxy acids, especially salicylic acid, are effective in exfoliating the lower stratum corneum layers, which results in smoother skin and increased cell division in the basal layer.

3. Unlike AHAs, BHAs exfoliate selectively and have been proven to be not only more effective, but also more gentle. In fact, 2 per cent of a beta-hydroxy acid can produce significantly better results than 6 per cent of an alpha-hydroxy acid.

4. Another benefit of the BHAs is that they can penetrate hair follicles because they are lipid-soluble, whereas alpha-hydroxy acids cannot. Therefore, BHAs can more effectively clean out pores by removing trapped impurities and dead cellular debris from them.

5. An interesting difference is that cells treated by BHAs appear younger looking under the microscope,[20] whereas they do not visibly change when treated with AHAs. Beta-lipohydroxy acid has been shown to mimic some of the stimulatory effects of retinoic acid on the epidermis and dermis, whereas glycolic acid does not.

6. Research results at this stage are more extensive with alpha-hydroxy acids than with beta-hydroxy acids. We will see a lot more interesting molecules for skin rejuvenation based on beta-hydroxy acids in preference to the alpha-hydroxy acids.

## REFERENCE

1. Van Scott, E.J. and R.J. Yu, *Hyperkeratinization, corneocyte cohesion, and alpha hydroxy acids.* J Am Acad Dermatol, 1984. **11**(5 Pt 1): p. 867-79.
2. Berardesca, E., et al., *Alpha hydroxy acids modulate stratum corneum barrier function.* Br J Dermatol, 1997. **137**(6): p. 934-8.
3. Windhager, K. and G. Plewig, *[Effects of peeling agents (resorcinol, crystalline sulfur, salicylic acid) on the epidermis of guinea pig (author's transl)].* Arch Dermatol Res, 1977. **259**(2): p. 187-98.
4. Yamamoto, Y., et al., *Effects of alpha-hydroxy acids on the human skin of Japanese subjects: the rationale for chemical peeling.* J Dermatol, 2006. **33**(1): p. 16-22.
5. Mishima, Y., *Molecular and biological control of melanogenesis through tyrosinase genes and intrinsic and extrinsic regulatory factors.* Pigment Cell Res, 1994. **7**(6): p. 376-87.
6. Yu, R.J. and E.J. Van Scott, *Alpha-hydroxyacids and carboxylic acids.* J Cosmet Dermatol, 2004. **3**(2): p. 76-87.

7. Rawlings, A.V., et al., *Effect of lactic acid isomers on keratinocyte ceramide synthesis, stratum corneum lipid levels and stratum corneum barrier function.* Arch Dermatol Res, 1996. **288**(7): p. 383-90.
8. Ditre, C.M., et al., *Effects of alpha-hydroxy acids on photoaged skin: a pilot clinical, histologic, and ultrastructural study.* J Am Acad Dermatol, 1996. **34**(2 Pt 1): p. 187-95.
9. Smith, W.P., *Epidermal and dermal effects of topical lactic acid.* J Am Acad Dermatol, 1996. **35**(3 Pt 1): p. 388-91.
10. Kockaert, M. and M. Neumann, *Systemic and topical drugs for ageing skin.* J Drugs Dermatol, 2003. **2**(4): p. 435-41.
11. Vilaplana, J., et al., *Clinical and non-invasive evaluation of 12% ammonium lactate emulsion for the treatment of dry skin in atopic and non-atopic subjects.* Acta Dermato-Venereologica, 1992. **72**(1): p. 28-33.
12. Huber, C. and E. Christophers, *"Keratolytic" effect of salicylic acid.* Arch Dermatol Res, 1977. **257**(3): p. 293-7.
13. Chiaretti, A., et al., *Salicylate intoxication using a skin ointment.* Acta Paediatrica, 1997. **86**(3): p. 330-1.
14. *Beta hydroxy acids: reversal of pre-mature ageing and treatment of adult acne.* 1998.
15. Pierard, G.E., et al., *Dermo-epidermal stimulation elicited by a beta-lipohydroxyacid: a comparison with salicylic acid and all-trans-retinoic acid.* Dermatology, 1997. **194**(4): p. 398-401.
16. Pierard, G.E., et al., *Comparative effects of retinoic acid, glycolic acid and a lipophilic derivative of salicylic acid on photodamaged epidermis.* Dermatology, 1999. **199**(1): p. 50-3.
17. *Safety assessment of Salicylic Acid, Butyloctyl Salicylate, Calcium Salicylate, C12-15 Alkyl Salicylate, Capryloyl Salicylic Acid, Hexyldodecyl Salicylate, Isocetyl Salicylate, Isodecyl Salicylate, Magnesium Salicylate, MEA-Salicylate, Ethylhexyl Salicylate, Potassium Salicylate, Methyl Salicylate, Myristyl Salicylate, Sodium Salicylate, TEA-Salicylate, and Tridecyl Salicylate.* Int J Toxicol, 2003. **22 Suppl 3**: p. 1-108.
18. Tsai, J., et al., *Distribution of salicylic acid in human stratum corneum following topical application in vivo: a comparison of six different formulations.* Int J Pharm, 1999. **188**(2): p. 145-53.
19. Lee, H.S. and I.H. Kim, *Salicylic acid peels for the treatment of acne vulgaris in Asian patients.* Dermatol Surg, 2003. **29**(12): p. 1196-9; discussion 1199.
20. Merinville, E., et al., *Three clinical studies showing the anti-ageing benefits of sodium salicylate in human skin.* J Cosmet Dermatol, 2010. **9**(3): p. 174-84.

# Chapter 15
# SKIN PEELING

## INTRODUCTION

Some people shudder when they think of skin peeling. First of all, they are worried about the pain, and secondly they are alarmed by stories of how one looks after the peel. Finally, they are terrified of their skin being permanently damaged after being treated with a peel.

Peeling is not new but goes back thousands of years. An ancient Egyptian papyrus describes a peel done for an old man to make him look younger, and it reports that it was very successful. (We should bear in mind that in those days an old man was something like 40 to 50 years old.) Peeling fell out of favour for the next 3000 years, but towards the end of the nineteenth century it started to become popular again, and by the 1950s heavy 'phenol' peeling was considered to be the only effective method to rejuvenate skin.

The frightening stories about peeling come from the experience with phenol. This heavy peel destroys the epidermis and the upper dermis almost right down to the hair roots and sweat glands. The surface epidermis must grow and cover the dermis again by activating the keratinocytes in the hair follicles and the sweat gland ducts. This takes at least ten days and the patient has literally got to be in hiding for that period. As a result of these deep phenol peels, the epidermis is very thin and flat and all the dermal papillae are destroyed and never return.

The skin becomes a bright pink colour that can last for up to a year, depending on the person's natural skin colour.

The dermis, however, is significantly scarred in the process and has very thick collagen deposition that smoothes out the skin and gives a younger appearance. The thick scar collagen in time makes the skin look abnormally waxy, white, and bloodless. However, in those days people were told they had to use make-up for the rest of their lives. They were also sun-sensitive for the rest of their lives. One positive spin-off was that they were less likely to develop a skin cancer in the peeled areas because it seems that all the cancer-prone cells were destroyed by the peeling process, and only less photodamaged cells remained. These cells became the new epidermis. The smoothing was at times marvellously impressive, and unfortunately this probably engendered the notion that the deeper, more aggressive the peel, the better the result. This concept ultimately led to carbon dioxide laser resurfacing as an alternative to deep phenol peeling. The belief that more aggressive peels give the best results unfortunately persists even today, and many thousands of people have suffered from that delusion. Predictably with such drastic and destructive changes, the skin takes on an unnatural look after a time, and not surprisingly phenol is being used much less frequently today. A serious risk in using phenol comes from its toxic effects on the heart leading to arrhythmia and even death.

Softer alternatives were explored, and one of them is salicylic acid for peeling, but, as mentioned in the previous chapter, high doses of topical salicylic acid can cause tinnitus, nausea, and vomiting. Hence, one has to be careful to use only the lower doses of salicylic acid, which in turn do not peel the skin very much. Tri-chloracetic acid was introduced because it has no dangerous side effects. While it is not as deep as phenol, it does give rise to a pretty deep destruction of the epidermis and upper dermis, which can still cause scarring. It takes on average about a week to recover from a deep tri-chloracetic acid peel.

AHA peels became popular because they are lighter and cause less damage. Glycolic gels of 50 to 70 per cent applied for about a minute and a half to two minutes became very popular in the USA. However, they were still judged according to the principle that one needed to cause destruction to get any effect.

### SENSIBLE 'COOL' PEELING

By carefully examining the clinical effects of various peels one can appreciate their shortcomings.

- There is an inherent problem in peeling because some people are very sensitive to even low doses of acid and will peel deeply, whereas others require much stronger acids to get a mild peel. This means that one has to individualize for every person, and experience reveals that the effects cannot easily be predicted.

*"The belief that more aggressive peels give the best results unfortunately persists even today, and many thousands of people have suffered from that delusion."*

> "Light repetitive peeling is an excellent way to treat acne, especially rampant acne, and one can treat both the face and the body this way."

❷ People applied liquids that dried quickly, making it impossible to see when specific areas had been treated twice, so over-treatment is a real problem with this type of peel.

A new view on peeling works on the concept that a five per cent acid applied to the skin for 15 to 20 minutes would cause less pain, but give the same results as a 35 per cent acid applied for one minute. This certainly reduces the pain, but by using simple principles inherent in normal physics one can get an even cooler peel: use low-concentration acids in a thick gel where the water component of the gel evaporates over several minutes. This will cool the surface through the process of heat loss by evaporation of water. Even though the acid is reacting into the skin, people are generally unaware of it. The application of a fan at the same time as performing the procedure will cool the skin even more by speeding up the evaporation process.

At the same time avoid damaging tissues any deeper than the epidermis. The old peels were designed to kill virtually the whole epidermis. The new concept of peeling works on the principle of stressing the basal keratinocytes but not killing them. Research has shown that short episodes of heat help to generate growth factors in yeast cells and short exposure to -12°C also causes release of growth factors[1]. When we do a peel we intend to rejuvenate the skin, and the best way to do that is to stimulate growth factors, and the new concept of 'cool' peeling works on the principle that if keratinocytes can be stressed sufficiently, they will release growth factors which will rejuvenate the skin, promote new collagen and elastin and stimulate the production of GAGs.[1] This is a regenerative mechanism rather than the destructive mechanism of traditional peeling techniques.

*Result achieved from light (5 per cent) trichloracetic acid cream peels done once every three weeks for eight months.*

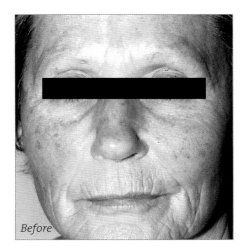
Before

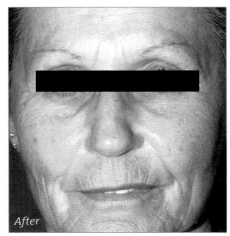
After

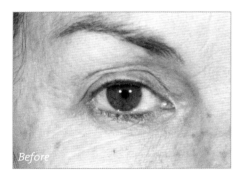
*Before*

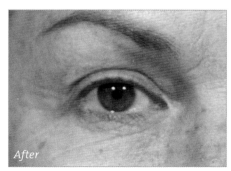
*After*

*Changes seen here are the result of using vitamin A and antioxidnat based cosmetic craams twice a day and having treatments with 2.5 per cent TCA cream applied for 20 minutes once a month. Notice the improvement of the thin upper eyelid skin especially on the medial side.*

In this way we avoid complications of dermal scarring and we promote the thickening of the epidermis that is, after all, our main defence against the environment. A distinct advantage of doing light peeling is that one can repeat the peel at fairly frequent intervals and then end up with the same result as a heavier peel but without going through the trauma or running the risk of unnecessary destruction of valuable skin layers and structures.

A most important point to remember is that no matter what type of peel one wishes to use, one should always prepare the skin with topical vitamins A, C, and E and antioxidants for at least three weeks prior to the treatment. Vitamin C will help to make more collagen and elastin and protect normal pigmentation production.

### COOL PEELING FOR ACNE

Light repetitive peeling is an excellent way to treat acne, especially rampant acne, and one can treat both the face and the body this way. The pictures below are an example of the effects of light peeling (TCA 2.5 to 5 per cent in a cream base) for acne initially done once a week and then once every two weeks. Eventually treatments were done once a month, followed by basic skin care with vitamins A and C and other antioxidants to keep the skin healthy and reduce sebaceous activity.

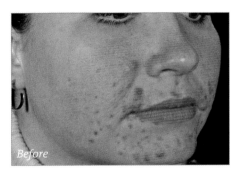
*Before*

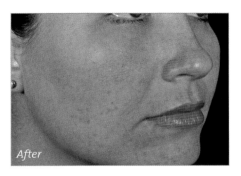
*After*

*Acne is very well treated by light peeling. In this case 2.5 to 5 per cent TCA in a cream formulation done on alternate days unless the acne settled. Continuous use of an oil containing retinyl palmitate, tocopherol and ascorbyl tetraisopalmitate has ensured success in treating the acne.*

Skin Peeling / 153

*Acne of the chest responds well to using retinyl palmitate oil daily and 5 per cent TCA once a week for three months*

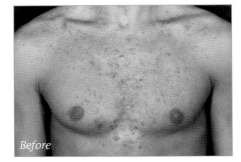
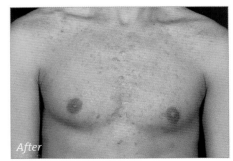

*Acne treated with retinyl palmitate creams twice daily in combination with a mild alpha hydroxyacid cream*

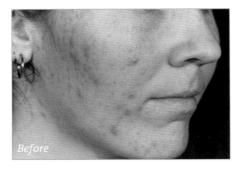
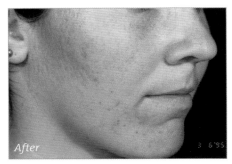

Acne of the chest causes a lot of psychological problems in young people, yet can surrender to regular peelings associated with topical vitamin A and cohort vitamins. This process may be improved by using neutralized lactic acid regularly as a skin toner with the topical vitamins.

In severe cases, acne can be very reliably and safely controlled by even daily cool peeling done with 2.5 per cent TCA or lactic acid applied for a maximum of ten minutes. Because of the short exposure of the low concentration of acid, one does not experience any exfoliation or sun sensitivity.

*Brian looked older than he should and looked younger after using retinyl palmitate and antioxidant creams twice a day and having TCA 5.0% peels once a month for six months*

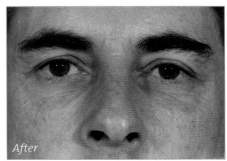

Before

After

*Changes seen after using retinyl palmitate and antioxidnat creams morning and evening for three months and then 4 days after five daily 'peels' with Lactic acid 5 per cent*

## COOL PEELING TO REJUVENATE SKIN

Brian looked too old for his age, so he was treated with a light serial peel once a month and kept on vitamins A and C and antioxidant skin care. The photographs show how he changed in eight months.

Cool peels may be done at home daily, in conjunction with weekly treatments by a professional to get the fastest results. This may be done with 2.5 or 5.0 per cent TCA or lactic acid. The regime starts with the professional applying the light acid for about 20 minutes in a thick layer. Then the patient is given 2.5 per cent TCA or lactic acid which they should apply daily for a maximum time of only ten minutes. Once again, it is important to continue using vitamins A and C and antioxidants twice a day. The picture shows the benefits achieved in just ten days.

## COOL PEELING TO TREAT SOLAR KERATOSES

The next pictures illustrate the value of light serial skin peeling as a method

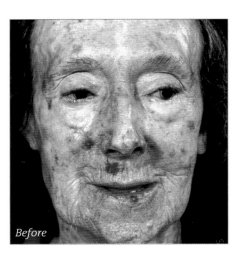
Before

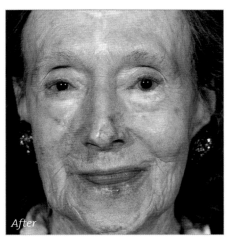
After

*87 year old with severe photodamage, actinic keratoses and numerous prior operations for basal cell carcinomata. Seen six months after starting rertinyl palmitate creams morning and evening and mild 5 per cent to maximum of 10 per cent TCA cream peels once a month for six months. Most of the keratoses have been cleared and the skin looks much healthier.*

*Exogenous ochronosis is a complication seen from high dose hydroquinone or prolonged low dose of hydroquinone topically. It is characterised by bluish black discolouration of tissues and cartilage or bone which show up under the microscope as yellowish granules. Acne and severe pigment blemishes are common. Even low doses of hydroqinone are best avoided despite their apparent value in treating pigmentation problems.*

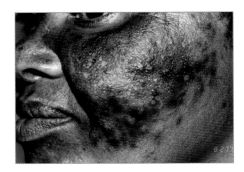

to reduce the risks of skin cancer. This lady of 86 had numerous skin cancers that had to be removed. A series of skin peels was done once a month and she used topical vitamins A and C and high-dose antioxidants twice a day. After six months she said the treatments had restored her dignity and self-confidence. She eventually lived to be over 100 years old and enjoyed soft facial skin.

Peeling has a role to play in the treatment of acne, wrinkles, and solar keratoses, but has only a limited role in the treatment of abnormal pigmentation. The pigmented cells in abnormal pigmentation are often situated in the deep dermis, deeper than we can reach with even 'deep' peeling. Consequently, although the abnormal pigmentation may get lighter or even disappear after a deep peel, it can reappear some years later and may

> *"After six months she said the treatments had restored her dignity and self-confidence."*

 **SAFETY RULES TO FOLLOW IN PEELING**

1. Never do heavy peeling. This is generally not safe for any skin.
2. Only do milder peels that can be repeated without any lasting stress to the skin.
3. Always prepare the skin for a peel by using topical vitamins A, C, and E and antioxidants for at least three weeks before a peel. It is even better if the skin has been prepared for three months or more. The antioxidants, like vitamins C and E, especially applied to skin when peeling is done, help to reduce the stress on the skin.
4. Avoid peeling in summer unless you are doing extremely light peels such as those that are done for acne. Those very light peels can safely be done all year round without causing problems.
5. Avoid bright sunlight on your skin for about three weeks after a 'deep' peel.
6. Always use a suitable sunscreen (not less than SPF 15 with UVA and UVB protection and antioxidants) with powerful inorganic component of light reflectors to protect fragile skin. Because the skin is in a fragile state, avoid using sunscreens with excessively high SPF and high concentrations of organic sunscreens, as these pose the danger of aggravating the free-radical challenge to the skin. Inorganic sunscreens may be used without restriction. Always reapply every hour and a half if you have had a peel, and wear a broad-brimmed hat that does not let sunlight through.

even be larger and darker than before. For abnormal pigmentation, one has to keep the epidermis thick, and one needs to treat the melanocytes with safe chemicals to change the amount of melanin formed and the way it is disseminated in the surrounding cells. One should avoid hydroquinone, which is a toxin to melanocytes and could cause ochronosis. It is difficult to get vitamin C easily and safely deep into the skin at adequate doses by simple topical application. This can, however, be done by using enhanced skin penetration techniques like cosmetic skin needling, and/or combined iontophoresis and sonophoresis.

These are the broad principles concerning peeling that can be done on most skin types. Although wonderful results can be achieved from repetitive serial light peeling, enhanced penetration techniques can lead to still further improvements.

## REFERENCE

1. Alappatt, C., et al., *Acute keratinocyte damage stimulates platelet-activating factor production*. Archives of dermatological research, 2000. **292**(5): p. 256-9.

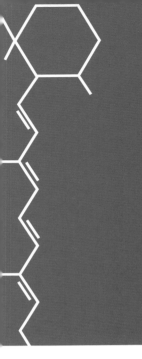

# Chapter 16
# WHEN SKIN CARE IS NOT ENOUGH. ENHANCED SKIN PENETRATION WITH IONTOPHORESIS AND SONOPHORESIS

## INTRODUCTION

The most exciting chapter in medical history is being written right now, as we are learning to manipulate the chemistry and physical properties of skin so that we can get important active molecules down to the cells that can make significant changes. For the first time in history we are able to treat the various aspects of photoageing by using active cosmetic creams that contain ingredients that physiologically minimize or even reverse the changes induced by excessive exposure to sunlight. The rejuvenation often exceeds what is achieved with complex medically oriented machines using lasers and thermal devices but produces healthier more radiant and natural skin.

The skin has always presented particular difficulties to the cosmetic scientist. The special properties of skin, specifically of the horny layer, make it very difficult to get most of the effective molecules through the epidermis into the lower parts of the skin where their effects can be expressed.[1, 2]

Until fairly recently it was thought that getting complex molecules through skin was impossible; however, as a result of enhanced skin penetration techniques, this has changed and some treatments may well be changing dramatically. For example, insulin is a very complex molecule that has been shown to penetrate the skin through the power of iontophoresis and low frequency sonophoresis.[3] This development opens up the possibility of giving diabetic patients insulin without any injections.

Ironically, the very improvement brought about by the active treatment with topical vitamin A further improves the barrier functions of the enhanced skin, thereby making it even more difficult to get the active molecules to target. We can call it the 'Vitamin A Paradox'. It is this very paradox that necessitates the use of specialized penetration techniques to ensure adequate flow of active molecules to the skin cells and components. Therefore, in order to get more significant results we need to get higher doses of the effective ingredients into the skin by enhancing penetration.

## PRINCIPLES OF ADVANCED PENETRATION ENHANCEMENT

The barrier function of skin should only be changed temporarily, with no damage to the structure of the skin. Iontophoresis and sonophoresis meet this requirement and have facilitated significant rejuvenation of skin. The simultaneous use of iontophoresis and sonophoresis is probably the most effective way of enhancing

### BASIC TECHNIQUES TO ENHANCE PENETRATION

There are a few different methods to facilitate penetration through skin.
1. We can manipulate formulations chemically to encourage better penetration. [4]
2. Peeling and microdermabrasion will enhance penetration but at a sacrifice of the normal function of the skin for about five to seven days until the stratum corneum has grown again. [5]
3. Iontophoresis (a mild direct-current electrical technology) has only recently been employed in a scientific method to enhance penetration of active cosmetic ingredients. [6] Iontophoresis does not make any permanent changes to the skin, so immediately after the treatment the skin returns to its normal state.
4. Ultrasound (1-3mHz) did not meet its promise of enhancing penetration and only enhanced penetration by insignificant degrees. [7]
5. In the 1990s that low-frequency ultrasound (about 20kHz) offered amazingly better penetration than high-frequency ultrasound.[8]
6. Another recent method to enhance penetration is to use micro-needles to penetrate the stratum corneum and painlessly bypass the skin's barrier. [9, 10] This has been applied to routine skin care by using a novel tool that makes temporary, self-sealing 'holes' in the stratum corneum.[11]

*Old acne scarring and photoageing treated with retinyl palmitate and acetate-based cosmetics with antioxidants and treatments to enhance penetration of vitamins A and C using low frequency sonophoresis combined with iontophoresis. One treatment done once a week for one year. This result far outranks results from fractional laser or thermal treatments, and in addition, she is left with normal smooth skin.*

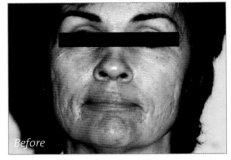
Before

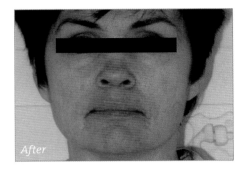
After

penetration that we know of today. The addition of micro-needling using needles that protrude only 0.1 to 0.2 mm prior to iontophoresis and sonophoresis gives an even greater chance of effecting skin rejuvenation. However, we have to adapt these techniques in order to use them in a professional cosmetic setting.

## IONTOPHORESIS

There has been a resurgence of interest in iontophoresis by the medical profession. We have known about galvanic current for about 200 years[12], and iontophoresis through skin was demonstrated almost 100 years ago[13]. Skin-care therapists have used galvanic currents to treat skin for about 70

*Pulsed iontophoresis is preferred because ionized of special molecules, e.g. ascorbic acid, occurs at the onset of the pulse. The positive ion in this case is attracted to the negative charge of the anode while the negatively charged ion is repelled away into the skin. After a while the ionisation ceases and the positive and negative ions join to form a molecule which is then ionised by another iontophoresis pulse. This continues till the current is stopped. It takes about 10 minutes or more before ions penetrate through the epidermis to the dermis.*

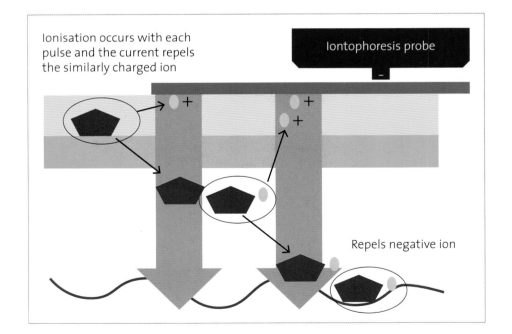

years, but they would 'activate' the skin on negative current for a few minutes, then massage the skin, and finally soothe it on positive for a few minutes to get a refreshed appearance. They were not actually using the power of iontophoresis to drive active chemicals into the skin.

Iontophoresis will only occur if the targeted active chemicals can ionize under physiological conditions. When a low-intensity current is applied to a molecule in solution with the potential to ionize, molecules will split into negatively and positively charged particles. Appropriately charged ions resulting from this ionizing process can be repelled deep into the skin and concentrated there by the weak currents used in iontophoresis.

We can use this phenomenon in active cosmetic skin treatments. If a water-based gel of an active cosmetic molecule that can ionize is placed on the skin and then a galvanic current is applied on the gel, the molecule will ionise into two ions or oppositely charged particles. When the active ion is positively charged, a positive current will drive the positively charged ion deeper into the skin, because like charges repel each other whereas negatively charged ions will be attracted to the electrode.

Should we mention that we *strongly do not condone* the abuse of animals? However, to understand iontophoresis better, we must go back to the very first demonstration of its power in 1908 by Le Duc[13]. In the experiment two rabbits were selected and one poison

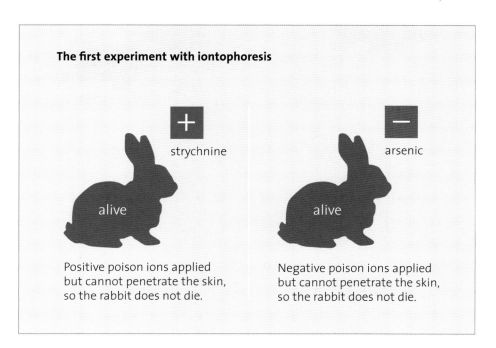

**The first experiment with iontophoresis**

strychnine

alive

Positive poison ions applied but cannot penetrate the skin, so the rabbit does not die.

arsenic

alive

Negative poison ions applied but cannot penetrate the skin, so the rabbit does not die.

*Killed rabbits*

*An illustration of Le Duc's experiment to show that the cathode current repels strychnine into the skin whereas the anodal current keeps the rabbit safe. On the other hand, the experiment with arsenic shows that positive cathodal current is not the cause of the first rabbit's death. In this case because arsenic is a negatively charged ion it is repelled by an anodal current and the rabbit dies whereas a cathodal current keeps the rabbit safe.*

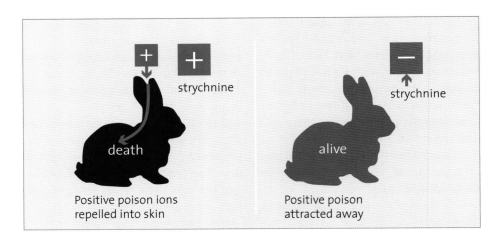

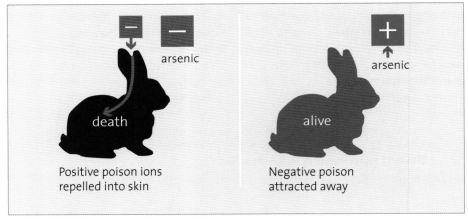

that had a positive charge was applied to one ear and a poison with a negative charge was applied to the other ear. If no current was used, the rabbits did not die. Then only one ear was treated with an electrical current. When the positively ionised poison was treated with a negative current, nothing happened. When positive current was used, the rabbit died. So the culprit was either positive current or the positively charged poison on the skin. The question was: could the positively charged poison ions move through the skin into the blood and kill the rabbit? The second rabbit then proved that it was the charged ions, because when a positive current was used, nothing happened, whereas when a negative current was used, the rabbit died because the negatively charged poison went into the skin. The only explanation was that the positive and negative ions had been repelled by their similar charges and had gone through the skin. This was a magnificent demonstration but unfortunately not enough people paid attention to it, and it was lumped together with things like hypnosis and ignored by the medical profession.

In recent years, research workers have looked again at iontophoresis and have tried to define its mysteries. As a result we have learned a great deal about it and today research workers believe that iontophoresis can very often be as effective as hypodermic injections into skin or muscles. Diabetics may one day routinely wear only a simple instrument like a watch and in that way first check the blood sugar levels and in response to the reading, dose themselves with enough insulin to keep healthy. Other powerful medicines can even be pushed more than 20 mm through the skin into the body by simple electrical currents[14]. Iontophoresis promises to become a major method for treating people without the use of injections and other invasive techniques. The research workers have realized that we have to know certain properties of ions, not merely that they can ionize.

It is now well known that the concentration of the ions is important. If there are too many ions then the 'pores' can become blocked. Therefore the strongest solution is not necessarily the best. We also know now that the properties of the current are important. Certain wave shapes and intermittent application of the current are better than continuous galvanic current.

## IMPORTANT PROVISIONS IN IONTOPHORESIS

We have to know certain specific properties of ions, not merely that molecules may be ionized. There are some very important rules that have to be followed when doing iontophoresis.

❶ The selected molecule must be ionized into positive and negative components and be maintained as ions during the treatment. One cannot do iontophoresis on chemicals that are not ionized during application of the current.

❷ The size of the ion is important.[15] For example, even though it may be possible to ionize a complex protein like collagen, the size of the important ion of collagen is simply so large that it cannot be transported through skin.

❸ There is a limit to the number of polar substances that can be used simultaneously. We believe that during iontophoresis 'pores' open up through membranes and the charged particles can move through them.[16, 17] If there are too many charged particles then the 'pores' may be blocked by the crowd of ions converging all at once.

❹ The ion must be water-soluble because electricity is conducted only through water and not through lipids. Therefore a water-based formulation or gel needs to be used as a carrier for the ions.

❺ The pH of the active gel is of fundamental importance. The right ingredient at the right concentration will not work properly if the pH is wrong. Each ion has its own ideal pH at which it will be ionized best. This fact is difficult to explain in simple terms, except to say that pH refers to the concentration of hydrogen (H+) atoms. These are ionized atoms and can interfere with an active, target molecule being ionized.

❻ The current used must be appropriate. When the active ion is negatively charged then you have to use a negative current, and if it is positively charged then only a positive current will get the active ions into the skin.

❼ The current used should be high enough to be effective, yet still be safe. The higher the current, the faster the ions will move.

❽ Intermittent current works better than continuous current because as the ion moves into the skin it will react with other chemicals, and that happens most easily in the period after the current[18]. Pulsed iontophoresis is also more comfortable.[19] Thereafter the molecule needs to be re-ionized by another 'blast' of current.

*Iontophoresis works by the developmenmt of 'pores' through the stratum corneum and epidermis. By 10 minutes enough 'pores' have been developed to allow detectable amounts of the treatment gel to go through the epidermis. By 30 minutes very little more penetrates, so for practical purposes, that is the maximum for iontophoresis treatments.*

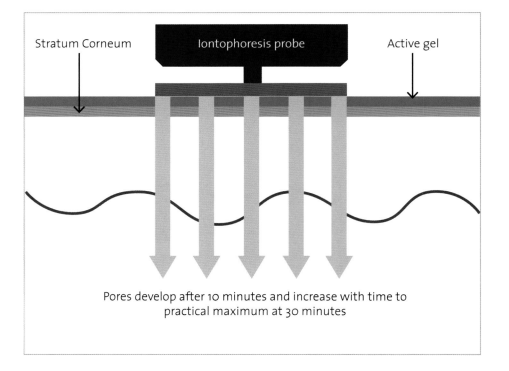

Pores develop after 10 minutes and increase with time to practical maximum at 30 minutes

9. The treatment period should be at least ten minutes and probably not longer than 60 minutes. Most of the ions pass through the skin at about eight to 15 minutes. Relatively fewer pass through after 30 minutes. We prefer to use a field of electrical charge rather than rollers, which produce rather localized and short-lasting effects. By using specialized gauze that retains moisture (Hazegauze®) or an alginate mask one can keep the current over the whole skin for a prolonged period.

10. It is possible, maybe even highly desirable, to treat skin with only one polarity in any one session. It is not necessary to immediately treat the skin with the opposite current as is usually advised by older skin care therapists, after doing the active treatment. If you do that, it will reverse the beneficial effects of the active treatment. If it is necessary to use two different currents, you should wait a short while between the sessions to allow the penetrated ions to stabilize and hopefully be absorbed into the skin.

11. Of course, the ion will only have positive effects if scientific research has proved that it is effective. Ionizing salt water into skin cannot have the same effects as ionizing a proven rejuvenating vitamin ion into the skin. This is where the manufacturers of active iontophoresis gels or serums should be requested to show research work to prove that their active ions do penetrate skin and make changes in the keratinocytes and fibroblasts to tighten and smoothen skin.

## ELECTRO-OSMOSIS

We are now also learning about electro-osmosis, which is closely allied to iontophoresis. In this case, ions move along the electrical 'concentration' gradient from a high current to a low current. [20, 21]

## ELECTROPORATION

Another intriguing phenomenon is electroporation. By this we mean the temporary 'holes' or passages created in membranes by short sharp electrical pulses that temporarily create "holes" in the stratum corneum and subsequently allow large or non-ionized molecules to pass through into the epidermis and through cellular membranes.[22] Electroporation uses higher currents than iontophoresis and every much shorter periods and can be quite uncomfortable.[23]

More recently, devices have appeared with micro needles for electroporation. [24] Studies have also suggested that electroporation may be a good way to get l-ascorbic acid into the skin and thereby hopefully tighten the skin.[25] However, it seems that this delivery system has not proved sufficient in clinical situations and more recent work focuses on combining electroporation with sonophoresis.[26]

In the field of delivering active molecules into the skin it appears from research that the ideal combination is iontophoresis in combination with low frequency sonophoresis. [27, 28]. The effects are even greater when combined with needling of the stratum corneum with a device with needles protruding between 0.1-0.2 mm.[29] However, there is an important distinction because needling cannot ensure penetration into cell walls, whereas iontophoresis and LFS does.

## TECHNICAL QUIRKS OF IONTOPHORESIS

Iontophoresis when correctly used is proving to be a powerful tool. However, it does have problems. First of all, it can be used only on molecules that can be ionized. Not all molecules can be ionized at physiological levels (e.g. retinyl palmitate) and some cannot be ionized at all. Secondly, if strong currents are used and the current is built up too quickly when using simple galvanic current, heat will result and a thermal burn may occur. Some very complex chemical changes occur at the site of the iontophoresis contact points. Strong alkalis or even strong acids may be formed in the ionization process that may damage the skin.

### EFFECTS OF TREATING SKIN WITH POSITIVE ACTIVE ELECTRODE:

1. Positive ions (cations) repel hydrogen ions and that causes the so-called acid effects on the active pole.
2. Electrons are stripped away from the positive electrode.
3. Chloride ions are attracted to the positive electrode and then combine with the electrons to become atomic chlorine, which can then act with water to become hydrochloric acid in the area of the positive electrode. This takes time and, of course, causes corrosion of the electrode.
4. A greater number of hydrogen (hydronium) ions move into the dermis and the dermal pH is lowered. We can use this fact for a gentle iontophoresis peel using relatively low doses of acid gels, e.g. lactic acid, glycolic acid, ascorbic acid etc.
5. Positive iontophoresis causes constriction of blood vessels, making the skin look pale. In the 'dark' pre-scientific times of cosmetic iontophoresis, this was considered as a sign of the skin being soothed which is why therapists tended to end a treatment with positive current to counteract the 'excess activity' caused by the negative current.
6. Positive (polarity) iontophoresis tends to be germicidal, which is another reason that it can be useful for treatment of acne.
7. Positive (polarity) iontophoresis reduces the excitability of the nerve fibres, which in turn has an analgesic effect. This apparent soothing of the skin is the reason that old-fashioned practitioners of iontophoresis insist on ending their treatments with a positive charge.

**EFFECTS OF TREATING ON NEGATIVE ACTIVE POLE:**

1. The negative electrode attracts electrons and hydrogen ions.
2. It also attracts sodium ions, which can take on an electron and become a sodium atom, which is then able to react with water to make sodium hydroxide, an alkali. This causes the alkali effects at the negative electrode and also causes corrosion of that electrode.
3. Surface oils and sebum are emulsified in the presence of a suitable conducting solution. This is called 'disencrustation' by older therapists. Sebum can be a problem because it can prevent the proper flow of current through the skin; that is the reason that the skin must be degreased before treatment starts.
4. Blood vessels are dilated and the skin becomes pinker. This was taken as a sign of activation and irritation of the skin, which then needed to be soothed with treatment with a positive current.
5. Nerve ends are stimulated in a mildly unpleasant way. This apparent irritation of the skin was always regarded with suspicion by the older therapists, who then reversed these changes by applying a positive current, but at the same time they also reversed any effects that they had created with the negative current.

Always remember that these chemical changes occur at both electrodes, and so if you are treating the face with negative current, the changes described above for negative current will happen on the face, whilst the changes described for positive current will be happening at the same time at the passive electrode.

Modern iontophoresis has been used to minimize wrinkles, eliminate pigment blemishes, soften scars, and normalize skin. However, this happens only when the gels used contain the right ingredients at the right pH with the right concentration and treated with the correct current properties for the correct amount of time. Lets examine these confining factors:

1. The ingredients that have been demonstrated with iontophoresis to minimize wrinkles and pigmentation, soften scars and normalize skin are vitamin A as retinol and retinoic acid. Retinyl palmitate is not ionized and any effects that have been shown may occur from passive flow with electro-osmosis. L-ascorbic acid or preferably, sodium ascorbyl phosphate[30], has been shown to tighten skin and reduce wrinkles and pigment marks[31] by inhibiting the action of tyrosinase. L-Ascorbic acid has also been shown to be effective in treating hypertrophic scars in unpublished studies by Fernandes. Peptides are also very suitable for iontophoresis but one has to know

*Clinical changes from iontophoresis alone. Stretch marks on the abdomen are treated with iontophoresis for 20 minutes of alternating retinol and ascorbic acid – 16 sessions in 16 weeks.*

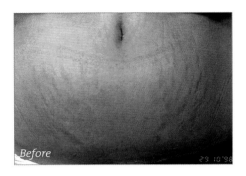
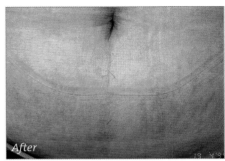

the polarity of the ion. Certain peptides have been shown to increase collagen and elastin formation[32] as well as glycosaminoglycans (e.g. Matrixyl 3000). Hyaluronic acid may be penetrated into skin by iontophoresis[33] and increase the intercellular fluid and thus smoothe out fine wrinkles.

❷ In general retinoic acid, retinol, ascorbic acid and sodium ascorbyl phosphate and hyaluronic acid require a negative charge in order to be repelled into the skin. The peptides will vary according to the dominant charge of the molecule. One cannot expect the same impressive results from using a positive current unless important changes have been done in the formulation of the product.

❸ Iontophoresis takes time to develop to change the resistance of the skin and this generally takes about 10 minutes.[34, 35] However, when one pulses the current then one radically changes the circumstances and the resistance is overcome faster. [36]

## RESULTS WHEN USING IONTOPHORESIS

It is believed that active ions that have been 'transported' are absorbed into the living cells, stored, and eventually produce changes. It is also believed that galvanic treatments can help to restore the normal electrical potential of ageing cells, thus improving their permeability.

## CONTRAINDICATIONS TO IONTOPHORESIS

❶ Infected skin disorders like acne or medical diseases, like herpes simplex or lupus erythematosus. You can treat acne patients and other conditions with pulsed current because the chance of causing a burn is unlikely[19], but you have to be extremely cautious when using a *continuous* galvanic current[37, 38].

❷ Open cuts, wounds and abrasions, because the area is inflamed it will have a higher moisture content and will therefore conduct electricity more actively. If necessary, cover these areas with a non-conductive gel like petroleum jelly. However, in certain cases doctors may elect to use iontophoresis of certain antibiotics into the wound to assist healing. The stronger the current, the greater a chance for developing a burn.

❸ Delicate, hypersensitive skin prone to allergies, and highly vascular (excessive telangiectasia/couperose) skin conditions should be treated only by experienced clinicians and only with pulsed iontophoresis.

❹ Clients suffering from cardiovascular diseases like angina pectoris or arrhythmias should first have the doctor's consent before treatment. A pacemaker or any other implanted electronic device is an absolute contraindication because this will cause interference with its activity and risk the patient's life.

❺ Epilepsy: electrical currents may precipitate an epileptic convulsion and a medical practitioner should be in attendance. It is best to use another form of penetration enhancement instead, such as low frequency sonophoresis.

❻ The first trimester of pregnancy: we do not know what electrical currents might do to foetal development, therefore delay any proposed treatments till after the first trimester. In the last month of pregnancy electrical currents might precipitate labour contractions, so it is advisable to avoid any treatments at that time.

❼ Sunburned or irritated skin conditions: electrical current will pass through the skin faster and more intensely and may be painful, and penetration through these areas may be deeper than anticipated. One also may be more likely to develop an electrical burn with continuous galvanic current.

❽ Excessive metallic fillings in the teeth and metal pins or plates in the face: continuous galvanic current may cause pain in the mouth or wherever metal plates etc. are.

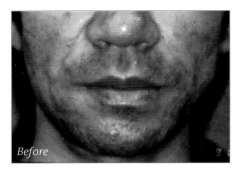

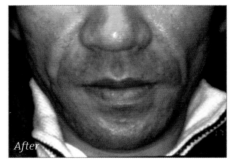

*Acne in a 35 year old man using retinyl palmitate and antioxidant cosmetics twice a day, and who had 24 iontophoresis treatments done twice a week, with a vitamin A gel for a total of 24 treatments.*

⑨ Sinusitis. If one uses a roller it can cause pain in the sinus area; however, with field and pulsed iontophoresis very much less discomfort is experienced.

⑩ Anaesthetic (numb) areas without sensation. The reason for being wary is that the client may not be able to feel an area of stinging when galvanic current is used and a burn could result when using continuous galvanic current. With pulsed iontophoresis this should not be a problem.

⑪ A highly nervous client. It is better to wait until the client really wants the treatment and has learned from others that it is a pleasant, relaxing experience.

Better results can follow if one does a complex treatment starting with a mini-peel and ending with a vitamin treatment. The mini-peel is done by gently 'iontophoresing' lactic acid into the skin, and then the skin is cooled and neutralized. The penetration properties of the skin are enhanced by a special alginate masque for doing iontophoresis of the vitamins A and C. This special facial procedure has proved to be very popular, and can even treat acne and at the same time reduce the scars. In addition the treatment is extremely comfortable.

However, iontophoresis has its own limitations. It can only be used if the targeted molecule can be ionized under physiological conditions. Larger molecules are also more difficult to get into the skin. That is where sonophoresis becomes important.

## LOW FREQUENCY SONOPHORESIS

If a chosen molecule will not ionize, you cannot use iontophoresis successfully, and microneedling of the skin or Low Frequency Sonophoresis (LFS) would be a better solution. LFS has now become the most promising way of enhancing penetration of molecules that do not ionize.

'Sonophoresis' means 'transport by the use of sound'. We have found that when sound is applied to skin, something like tunnels appear that allow molecules through the epidermis into the dermis. These passages seem to be the same as the holes that occur with iontophoresis but depending on the sound used, may be larger.

Many people are aware of ultrasound, which is the sound above the highest note that we can hear. Ultrasound is used in diagnostic investigations of the body, but those are very different sounds from the ones that we are interested in, which are only the much lower-frequency sounds, just above human hearing. These are the sounds that dolphins make when they swim. You may have heard that being in a pool with dolphins can treat people with depression and other psychological problems. The dolphins send out sound waves that penetrate human skin and allow water to enter the skin.

The power of these sound waves is tremendous, and we prefer to use them as an advanced penetration technique. It is estimated that using sonophoresis under the best conditions, we can get up to 30 or 40 times better penetration than simple topical application. The problem is that we can only treat very limited areas. However, it is an ideal way to treat smaller areas (such as abnormal pigmentation) that do not respond to other treatments.

The original research by Mitragotri et al. in this field is fascinating. At first it seems unbelievable. This is a technique that promises amazingly enhanced penetration of molecules, and we need a machine that would allow these powerful sounds to work for us in skin treatments. That proved to be a major obstacle because no one seemed able to produce a suitable machine until an amazing man in Japan who was the doyen of instrument manufacturers for beauty therapists in Japan took on the challenge. Eventually he was able to make a machine to our specifications that could deliver the best sonophoresis sound waves to the skin, and a whole new field in skin care was opened. This machine is patented to Dr Des Fernandes and is only supplied to therapists using Environ skin care because that is the only product that has been tested for safety.

## TOPICAL DELIVERY OF COSMETIC INGREDIENTS BY SONOPHORESIS

In the mid 1900s ultrasound was introduced to enhance penetration through the skin.[39-41] Therapeutic ultrasound (frequency: 1-3 MHz and intensity: 0-2 W/cm²) is the best-known form of ultrasound used to enhance transdermal drug transport. However, therapeutic ultrasound typically enhances penetration by less than 20 per cent. In truth the effects of high frequency ultrasound can be likened to the effects seen from occluding the treated area with a plastic wrap for 20 minutes.

Ultrasound alters pore pathways through the skin by two possible mechanisms: (a) enlarging the skin effective pore radii, or (b) creating more effective pores (radii of about 28 +/- 13 A)[42, 43] and/or making the pores less tortuous.[43]

However, recent research by Mitragotri et al. has shown that LFS (Low Frequency Sonophoresis) at about 20 kHz (125 mW/cm2, 100 msec pulses applied every second) induces transdermal transport enhancements of up to 1000 times higher than those induced by therapeutic ultrasound.[8] Since cavitation effects vary inversely with ultrasound frequency,[44] they hypothesized that ultrasound at lower frequencies should induce more cavitation and be more effective than therapeutic ultrasound in enhancing transdermal drug transport. This is low frequency sonophoresis (LFS).

The induction of cavitation will not automatically occur at the onset of phonophoresis (sound transport in general) unless a particular energy density (which equates to loudness) has been crossed: this is the threshold level. As you increase the frequency of the sound, you also have to increase the energy of the sound in order to reach the threshold to generate cavitation. You have to use about 100 times greater energy density to induce cavitation at 150 kHz as compared to 20 kHz, yet there would be even more cavitation at 20 kHz by using the same energy level, because once the threshold value is crossed, the enhancement of penetration becomes directly proportional to the ultrasound energy density. At this stage we know only the threshold for full thickness pig skin (about 222 J/cm$^2$),[45] but we do not know the threshold for human skin. In addition, it has been shown that low frequencies (approximately 20 kHz) have a more localized effect in increasing penetration as compared to a more dispersed effect seen at higher frequencies (for example 58 kHz).[45, 46]

## CAVITATION AND SONOCHEMISTRY

Cavitation bubbles from therapeutic ultrasound implode violently in less than a microsecond, and the contents can be heated to 5,500 degrees Celsius.[46] For this reason the ultrasound probe has to be kept moving over the skin to avoid a burn. On the other hand, with LFS the skin is unaffected according to both histology and electron microscopy studies.[47] Human skin samples exposed to 20 kHz at intensities lower than 2.5 W/cm$^2$ showed no modification at all.[48] From that we can conclude that not only is LFS extremely effective to enhance penetration through skin, it is also extremely safe. Furthermore, the evidence clearly shows that high-frequency ultrasound (for example, 1.1 MHz) is not only very inefficient at promoting penetration through skin, but it is also potentially harmful.

LFS (Low Frequency Sonophoresis) is a completely new technique and has been used primarily for medical indications in circumstances that are very different from those found in the average skin-care clinic. LFS was designed to deliver medications with such complex molecules as insulin through small areas of skin, thereby removing the necessity for administering injections. This is in sharp contrast to its use in skin care where larger areas need to be treated. The principles of LFS have therefore been adapted by Fernandes so that this valuable technique can be applied to the skin-care clinic. This has been patented in Japan and only Environ Skin Care has been permitted to use this machine.

The main effects of LFS occur in the stratum corneum but are not confined to it. The lamellar bodies and the lipid bi-layers that surround the corneocytes seem to be the most vulnerable to sonophoresis. It is said in theory that LFS causes cavitation with the oscillation of the sound and the collapse of gaseous cavities. The oscillations of the cavitation bubbles may cause increased water penetration into the lipid bi-layers and thereby create water channels through which water-soluble ingredients in solution may pass through the stratum corneum.

These disruptions are called lacunae, or literally 'small lakes'. Under ordinary conditions, lacunar domains are few and far between. With LFS the lacunae become numerous and may even become continuous. This creates a pore pathway for penetration of polarized- (ions) and non-polar molecules across the horny layer[49] which are virtually as large as the stratum corneum is thick.

These pores are considerably larger than the pores induced by iontophoresis, which are 1000 times smaller (about 0.0038 μm in diameter). This cavitation occurs both in the layers around the cell as well as on the cell walls and causes aqueous channels into the cells themselves.[8] With LFS the transport channel pathway is significantly shortened as compared to the normal tortuous pathway in between cells.

At 168 kHz continuous ultrasound the pores are about 20 microns: large enough to allow the transdermal passage of high-molecular-weight molecules that normally cannot penetrate skin.[50] At 20 kHz with lower energy densities than at 168 kHz, you can anticipate that the pores will be just as large and as effective. It has been found that molecules that normally diffuse passively through the skin at a relatively slow rate seem to be preferentially enhanced by sonophoresis.[51]

Pulsed (discontinuous) LFS may give greater enhancement of penetration of molecules than continuous LFS. Skin histopathology and permeation experiments demonstrate that both disordering of stratum corneum and convective flow were the main reasons for enhanced penetration. It is possible that the increased flow (permeation) of molecules occurs through both hair follicles and sweat ducts.[52]

A short application of LFS (which has been shown clinically to be about 15 seconds in unpublished studies by Fernandes) rapidly increased skin permeability and lasted for at least three hours.[53] During this three-hour period, drugs may be administered and successfully transported into the skin.[54] That is why in relation to skin care, one has to be particularly selective about which products one applies to the skin after an LFS treatment session.

## LFS (LOW FREQUENCY SONOPHORESIS) MACHINES

In essence, a sonophoresis machine should have a 'sonicator', or specialized vibrating unit, which is the component that creates the sound. The sonicator is connected to an applicator (probe) that transmits the sound to the skin. The applicator should have a flat surface for making the most efficient contact with skin. The applicator should be a disc to transmit the sound most efficiently. Because of the difference in energy densities, the size of the applicator is much smaller for LFS than for high frequency ultrasound. Thin-edged scraping type of applicators are not suitable for treating large areas of skin to enhance cavitation and penetration through the epidermis.

The SonoPrep® device manufactured by Sontra (Sontra Medical Corporation,

### TO MAXIMIZE THE CLINICAL EFFECTIVENESS OF LFS

1. The topical drug (both the drug and the carrying agent) should transmit the sound, in other words allow the sound to penetrate through to the skin. Basically this means that water-soluble molecules should be used for the treatment when sonophoresis is used.
2. Sodium lauryl sulphate (SLS), a well-known surfactant, can enhance LFS by lowering the threshold energy density by about ten times. The synergistic effect of SLS and LFS on transdermal transport increases linearly with SLS concentration.[56]
3. Heating the skin and moistening it with steam, or even just shaving, will enhance penetration.[57] For this reason, cleanse the skin thoroughly prior to LFS with a cleanser containing SLS (or Sodium Laureth Sulphate – SLES) then steam thoroughly to heat and moisten the skin.
4. The treated skin should be occluded to prevents escape of moisture and to take advantage of the enhanced penetration after the treatment.[58] A good alginate masque would be excellent here.

LFS has been shown to increase the permeability of human skin by several orders of magnitude,[29, 59] thus making it possible to deliver doses of proteins such as insulin, interferon gamma, and erythropoietin. The molecules of vitamins A, C, E, beta-carotene, B12, smaller peptides, and other active cosmetic ingredients are tiny in comparison to these large proteins. LFS may also enhance the penetration of growth factors and other important modulators of skin cells.[29] In fact, LFS opens up a whole new dimension in the rehabilitation of photoaged or diseased skin.

10 Forge Parkway, Franklin, MA 02038 USA. www.sontra.com) applies LFS energy (55 kHz) to the skin for 15 seconds (average) through a liquid coupling medium to create cavitation bubbles that expand and contract in the coupling medium. LFS cavitation creates reversible micro channels about the size of a human hair in the skin through which even large molecules can also be delivered. As a result local anaesthetic cream can numb the skin within five minutes instead of an hour. This device treats only a small area (0.8cm$^2$) and therefore is not suitable for the delivery of active cosmetic ingredients into skin.

When treating areas for cosmetic reasons, you have to adapt the principles to treat much larger areas. At the time of writing, only one machine (the Environ Ionzyme DF machine) allows skin-care therapists to employ LFS in an effective and practical manner. This machine has been used since 1997 for LFS with or without pulsed iontophoresis and has been tested at Queensland University where it was shown to enhance penetration tenfold (unpublished[55]). Because larger areas are being treated one cannot expect the same speed and degree of penetration enhancement as is found in the laboratory or with the Sonoprep. However, LFS will still enhance penetration more than any other device available to the skin-care therapist, and one can achieve demonstrable effects from 15 seconds application.

Fernandes has tested the application of a neuro-peptide complex by applying LFS with positive charge using the Environ DF II machine. He found that if the probe was applied for five to ten seconds per area, then no effect could be demonstrated. However, at 15 seconds per area, inhibition of muscular activity could be induced. Twenty seconds gave a more pronounced effect. At the time of writing this has not been published.

## COMPLICATIONS OF LFS

Low-frequency ultrasound at low intensities appears safe for use to enhance the topical delivery of active molecules. Researchers found only minimal urticarial reactions in dogs, but when they used higher-intensity and higher frequency sound it resulted in second-degree burns, most likely attributable to localized heating.[60] Fortunately, by using LFS in the region of 20 kHz one avoids these complications. Low-frequency ultrasound at low intensities is safe for use to enhance the topical delivery of active cosmetics.

## THE COMBINATION OF IONTOPHORESIS AND LFS

Iontophoresis is well established as a technique to enhance percutaneous (through the skin) penetration of active cosmetic molecules. When low-frequency sonophoresis and iontophoresis are done simultaneously, one can expect the following benefits:

❶ Higher transdermal fluxes of the applied active cosmetic because of the combined cavitation induced by LFS and iontophoresis in combination.[28]

❷ Because cavitation occurs within 15 seconds this will allow rapid transdermal passage of the active cosmetic. When using iontophoresis alone, there is a lag phase lasting about five to ten minutes before penetration of the skin accelerates. The combination of iontophoresis and sonophoresis causes cavitation to occur even earlier.[27, 28, 61, 62]

❸ Iontophoresis lowers the threshold of sound-energy densities required to cause cavitation.[60] This further increases the effectiveness of the cavitation produced by LFS and makes the treatment even safer but still highly effective at lower sound-energy intensities.

❹ The combination may enhance the penetration of molecules that do not ionize.

*The power of LFS combined with iontophoresis in conjunction with a triple peptide complex, specially designed for use with medical skin needling. The results after 8 years are very evident, showing skin tightening, reduction of pigmentary blemishes with beautiful smooth skin.*

Before

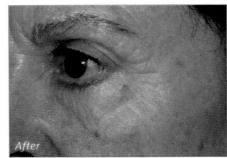
After

❺ Larger ions will be able to penetrate through the stratum corneum as compared to iontophoresis without LFS.

❻ Still further enhancement can be obtained by preparing the skin using a combination of chemical enhancers like ethanol, SLES, linoleic acid, propylene glycol DMSO, or other similar enhancers.[21]

Cavitation may play a two-fold role in enhancing the effect of iontophoresis:

**(A)** as described above, sonic oscillations induce cavitation and partial structural disorder in the skin's lipid bi-layers. Since the electrical resistance of the disordered bi-layers is likely to be smaller than that of the normal ones, the applied electric field may be as much as 12 per cent more efficient;

**(B)** The oscillations of cavitation bubbles may also induce convection across the skin and facilitate the flow of ions through the skin.[61] Convection in this sense means the funnelled concentration of the flow of molecules.

## THE TECHNIQUE FOR THE SKIN-CARE THERAPIST

When using a sonoprobe we have to remember that sound travels better

*Typical pigmentation seen in Japanese people. This was treated with maximum doses of vitamin A,C and also niacinamide and undecylenoyl phaynlalanine and LFS of ammonium lactate and vitamin A.*

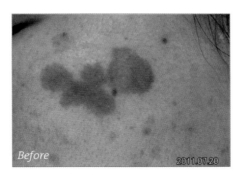
Before 2011.07.20

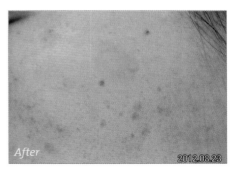
After 2012.08.23

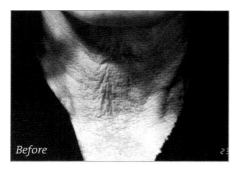
Before

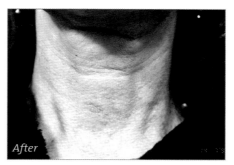
After

*This patient had a face and neck lift one year before and presents with elastosis and inelasticity of her lower neck skin. Twenty-four treatments of LFS and iontophoresis done in four months made a remarkable difference. Retinol was used, alternating each treatment with ascorbic acid.*

through water than through air. Because it is impractical to build up effective dams filled with hydrogel in the same way as the original experiments were done to show the power of LFS, we have to hold the sonicator as close to the skin as possible, with a layer of hydrogel between it and the skin. This layer can be even less than 1 mm thick and will conduct the sound to the skin surface provided that there are no entrapped air bubbles in the gel. In order to ensure efficient transmission of the sound, there should be no air space between the applicator and the skin.

The active product should be applied and this is preferably in a hydrogel formulation or may be covered by a thin layer of hydrogel. The sono-probe should be kept stationary for a minimum of 15 seconds then moved slowly to the next position but there has to be sufficient hydrogel to glide over the skin, cushioned by the hydrogel. The sound will then be fairly efficiently transmitted from the sonicator to the skin. Only relatively small areas of skin the size of the sonicator probe can be treated at a time, but over about eight minutes a relatively large area e.g. the forehead can be treated. The whole face can be treated in this manner, or one could treat selected areas. The temptation is to move the probe around on the skin but then this means that the skin is subjected to LFS for only a fraction of the time it takes to cause cavitation.

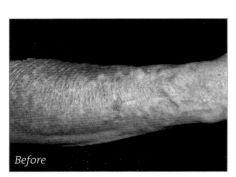
Before

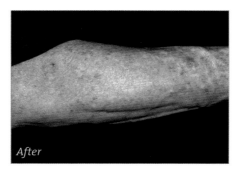
After

*Extreme photo-damage of the arms treated with LFS and iontophoresis of retinol to promote normalisation of the skin. Result after 24 sessions in three months.*

A pilot study has shown that pre-treating skin with sonophoresis alone for ten minutes prior to conducting an iontophoresis experiment resulted in a tenfold (1000 per cent) increase in permeation through the skin.[55] By employing LFS and iontophoresis simultaneously, even better penetration should ensue. This enables skin-care therapists to use cosmetic levels of retinol to achieve smoothing of skin normally seen only with high doses of retinoic acid.

LFS can deliver large molecules such as insulin, interferon gamma, and erythropoietin into the skin. The molecules of vitamins A, C, E, beta-carotene, B12, peptides and other cosmetic active ingredients, are tiny in comparison to these large proteins. In future LFS may also enhance the penetration of hormones and other important modulators of skin cells. In fact, LFS opens up a whole new dimension in the rehabilitation of photoaged or diseased skin.

One has to carefully weigh out the risk-benefit in trying to enhance the penetration of growth factors. We have inherited ideal concentrations of growth factors for various situations and often growth factors work in concert with each other and these ratios are not yet known. We have to question if it is safe to enhance the penetration of one or more growth factors. The concentration of growth factors in colostrum seems to be safe and studies over more than ten years indicate that they are safe even when enhanced with sonophoresis.

## CLINICAL RESULTS OF USING IONTOPHORESIS AND LFS

There is a need for enhanced penetration in treating photoageing and related conditions. Iontophoresis and LFS, individually and especially when combined, have opened up opportunities for skin-care therapists to treat photoageing and achieve results that have never before been possible. They are safe and more effective than any other modality available to the skin care therapist and can give results, comparable and even superior to, results achieved by scarring skin with fractional lasers or heat treatments. Every skin care therapist who desires to make significant changes by using only cosmetics has to employ the power of iontophoresis and LFS.

## REFERENCES

1. Bouwstra, J., et al., *New aspects of the skin barrier organization.* Skin Pharmacol Appl Skin Physiol, 2001. **14 Suppl 1**: p. 52-62.
2. Bouwstra, J.A. and M. Ponec, The skin barrier in healthy and diseased state. Biochim Biophys Acta, 2006. **1758**(12): p. 2080-95.
3. Ogura, M., S. Paliwal, and S. Mitragotri, *Low-frequency sonophoresis: current status and future prospects.* Adv Drug Deliv Rev, 2008. **60**(10): p. 1218-23.
4. Barry, B.W., *Novel mechanisms and devices to enable successful transdermal drug delivery.* Eur J Pharm Sci, 2001. **14**(2): p. 101-14.
5. Karimipour, D.J., G. Karimipour, and J.S. Orringer, *Microdermabrasion: an evidence-based review.* Plast Reconstr Surg, 2010. **125**(1): p. 372-7.
6. Fernandes , D.B., *Evolution of Cosmeceuticals and Their Application to Skin Disorders, Including Aging and Blemishes*, in Dermatological and

Cosmeceutical Development: Absorption Efficacy and Toxicity 2007. p. 45-60.

7. Byl, N.N., *The use of ultrasound as an enhancer for transcutaneous drug delivery: phonophoresis.* Phys Ther, 1995. **75**(6): p. 539-53.
8. Mitragotri, S., D. Blankschtein, and R. Langer, *Transdermal drug delivery using low-frequency sonophoresis.* Pharm Res, 1996. **13**(3): p. 411-20.
9. Prausnitz, M., M. Allen, and S. Davis, *Microfabricated Microneedles for transdermal drug delivery.* Perspectivesd in Percutaneous Penetration, 2000. **7a**: p. 4.
10. Henry, S., et al., *Microfabricated microneedles: a novel approach to transdermal drug delivery.* J Pharm Sci, 1998. **87**(8): p. 922-5.
11. Fernandes, D., *Percutaneous collagen induction: an alternative to laser resurfacing.* Aesthet Surg J, 2002. **22**(3): p. 307-9.
12. Helmstadter, A., *The history of electrically-assisted transdermal drug delivery ("iontophoresis").* Die Pharmazie, 2001. **56**(7): p. 583-7.
13. Chien YW, B.A., *Iontophoretic (transdermal delivery of drugs: overview of historical development.* J Pharm Sci, 1989. **78**(5): p. 353-4.
14. Scott, E.R., et al., *Transport of ionic species in skin: contribution of pores to the overall skin conductance.* Pharm Res, 1993. **10**(12): p. 1699-709.
15. Mudry, B., et al., *Quantitative structure-permeation relationship for iontophoretic transport across the skin.* J Control Release, 2007. **122**(2): p. 165-72.
16. Ferry, L.L., *Theoretical model of iontophoresis utilized in transdermal drug delivery.* Pharm Acta Helv, 1995. **70**(4): p. 279-87.
17. Uitto, O.D. and H.S. White, *Electroosmotic pore transport in human skin.* Pharm Res, 2003. **20**(4): p. 646-52.
18. Johnson, P.G., et al., *A pulsed electric field enhances cutaneous delivery of methylene blue in excised full-thickness porcine skin.* The Journal of investigative dermatology, 1998. **111**(3): p. 457-63.
19. Reinauer, S., et al., *[Pulsed direct current iontophoresis as a possible new treatment for hyperhidrosis].* Hautarzt, 1995. **46**(8): p. 543-7.
20. Grimnes, S., *Skin impedance and electro-osmosis in the human epidermis.* Med Biol Eng Comput, 1983. **21**(6): p. 739-49.
21. Batheja, P., R. Thakur, and B. Michniak, *Transdermal iontophoresis.* Expert Opin Drug Deliv, 2006. **3**(1): p. 127-38.
22. Banga, A.K. and M.R. Prausnitz, *Assessing the potential of skin electroporation for the delivery of protein- and gene-based drugs.* Trends Biotechnol, 1998. **16**(10): p. 408-12.
23. Banga, A.K., S. Bose, and T.K. Ghosh, *Iontophoresis and electroporation: comparisons and contrasts.* Int J Pharm, 1999. **179**(1): p. 1-19.
24. Choi, S.O., et al., *An electrically active microneedle array for electroporation.* Biomed Microdevices, 2010. **12**(2): p. 263-73.
25. Zhang, L., et al., *Electroporation-mediated topical delivery of vitamin C for cosmetic applications.* Bioelectrochem Bioenerg, 1999. **48**(2): p. 453-61.
26. Wang, Y., et al., *Transdermal iontophoresis: combination strategies to improve transdermal iontophoretic drug delivery.* European journal of pharmaceutics and biopharmaceutics : official journal of Arbeitsgemeinschaft fur Pharmazeutische Verfahrenstechnik e.V, 2005. **60**(2): p. 179-91.
27. Fang, J.Y., et al., *Transdermal iontophoresis of sodium nonivamide acetate. V. Combined effect of physical enhancement methods.* International journal of pharmaceutics, 2002. **235**(1-2): p. 95-105.
28. Mitragotri, S. and J. Kost, *Low-frequency sonophoresis: a review.* Adv Drug Deliv Rev, 2004. **56**(5): p. 589-601.
29. Kalluri, H. and A.K. Banga, *Transdermal delivery of proteins.* AAPS PharmSciTech, 2011. **12**(1): p. 431-41.
30. Marra, F., et al., *In vitro evaluation of the effect of electrotreatment on skin permeability.* J Cosmet Dermatol, 2008. **7**(2): p. 105-11.
31. Huh, C.H., et al., *A randomized, double-blind, placebo-controlled trial of vitamin C iontophoresis in melasma.* Dermatology, 2003. **206**(4): p. 316-20.
32. Pickart, L., *The human tri-peptide GHK and tissue remodeling.* J Biomater Sci Polym Ed, 2008. **19**(8): p. 969-88.
33. Pacini, S., et al., *Pulsed current iontophoresis of hyaluronic acid in living rat skin.* Journal of dermatological science, 2006. **44**(3): p. 169-71.
34. Theiss, U., I. Kuhn, and P.W. Lucker, *Iontophoresis--is there a future for clinical application?* Methods Find Exp Clin Pharmacol, 1991. **13**(5): p. 353-9.
35. Akomeah, F.K., G.P. Martin, and M.B. Brown, *Short-term iontophoretic and post-iontophoretic transport of model penetrants*

across excised human epidermis. Int J Pharm, 2009. **367**(1-2): p. 162-8.
36. Kanebako, M., T. Inagi, and K. Takayama, *Evaluation of skin barrier function using direct current II: effects of duty cycle, waveform, frequency and mode*. Biological & pharmaceutical bulletin, 2002. **25**(12): p. 1623-8.
37. Costello, C.T. and A.H. Jeske, *Iontophoresis: applications in transdermal medication delivery*. Phys Ther, 1995. **75**(6): p. 554-63.
38. Lambert, D., et al., *[Accidents caused by iontophoresis]*. Ann Dermatol Venereol, 1993. **120**(12): p. 907-8.
39. Levy, D., et al., *Effect of ultrasound on transdermal drug delivery to rats and guinea pigs*. The Journal of clinical investigation, 1989. **83**(6): p. 2074-8.
40. Bommannan, D., et al., *Sonophoresis. I. The use of high-frequency ultrasound to enhance transdermal drug delivery*. Pharmaceutical research, 1992. **9**(4): p. 559-64.
41. Birnbaum, Y. and A. Battler, *Augmentation of reperfusion by noninvasive, transcutaneous delivery of low-frequency, high-intensity ultrasound*. International journal of cardiovascular interventions, 2000. **3**(3): p. 137-141.
42. Tang, H., et al., *Theoretical description of transdermal transport of hydrophilic permeants: application to low-frequency sonophoresis*. J Pharm Sci, 2001. **90**(5): p. 545-68.
43. Tezel, A., A. Sens, and S. Mitragotri, *Description of transdermal transport of hydrophilic solutes during low-frequency sonophoresis based on a modified porous pathway model*. J Pharm Sci, 2003. **92**(2): p. 381-93.
44. Gaertner, W. and *Frequency dependence of ultrasonic cavitation*. J. Acoust. Soc. Am, 1954. **26**: p. 977-80.
45. Mitragotri, S., et al., *Determination of threshold energy dose for ultrasound-induced transdermal drug transport*. J Control Release, 2000. **63**(1-2): p. 41-52.
46. Suslick, K.S., *Sonochemistry*. Science, 1990. **247**(4949): p. 1439-45.
47. Boucaud, A., et al., *In vitro study of low-frequency ultrasound-enhanced transdermal transport of fentanyl and caffeine across human and hairless rat skin*. Int J Pharm, 2001 **228**(1-2): p. 69-77.
48. Boucaud, A., et al., *Clinical, histologic, and electron microscopy study of skin exposed to low-frequency ultrasound*. Anat Rec, 2001. **264**(1): p. 114-9.
49. Menon, G.K. and P.M. Elias, *Morphologic basis for a pore-pathway in mammalian stratum corneum*. Skin Pharmacol, 1997. **10**(5-6): p. 235-46.
50. Wu, J., et al., *Defects generated in human stratum corneum specimens by ultrasound*. Ultrasound Med Biol, 1998. **24**(5): p. 705-10.
51. Mitragotri, S., et al., *A mechanistic study of ultrasonically-enhanced transdermal drug delivery*. J Pharm Sci, 1995. **84**(6): p. 697-706.
52. Fang, J., et al., *Effect of low frequency ultrasound on the in vitro percutaneous absorption of clobetasol 17-propionate*. Int J Pharm, 1999. **191**(1): p. 33-42.
53. Mitragotri, S., et al., *Synergistic effect of low-frequency ultrasound and sodium lauryl sulfate on transdermal transport*. J Pharm Sci, 2000. **89**(7): p. 892-900.
54. Mitragotri, S. and J. Kost, *Low-frequency sonophoresis: a noninvasive method of drug delivery and diagnostics*. Biotechnol Prog, 2000. **16**(3): p. 488-92.
55. Bonnefoy, J., et al., *Interim report Ionzyme DF Sonophoresis/Iontophoresis Studies*. (unpublished). 2006(June-August).
56. Benfeldt, E., J. Serup, and T. Menne, *Effect of barrier perturbation on cutaneous salicylic acid penetration in human skin: in vivo pharmacokinetics using microdialysis and noninvasive quantification of barrier function*. Br J Dermatol, 1999. **140**(4): p. 739-48.
57. Clarys, P., et al., *In vitro percutaneous penetration through hairless rat skin: influence of temperature, vehicle and penetration enhancers*. Eur J Pharm Biopharm, 1998. **46**(3): p. 279-83.
58. Byl, N., *The use of ultrasound as an enhancer for transcutaneous drug delivery: phonophoresis*. Phys Ther, 1995. **75**(6): p. 539-53.
59. Priborsky, J. and E. Muhlbachova, *Evaluation of in-vitro percutaneous absorption across human skin and in animal models*. J Pharm Pharmacol, 1990. **42**(7): p. 468-72.
60. Singer, A.J., et al., *Low-frequency sonophoresis: pathologic and thermal effects in dogs*. Acad Emerg Med, 1998. **5**(1): p. 35-40.
61. Hikima, T., et al., *Mechanisms of synergistic skin penetration by sonophoresis and iontophoresis*. Biological & pharmaceutical bulletin, 2009. **32**(5): p. 905-9.
62. Katikaneni, S., et al., *Transdermal delivery of a approximately 13 kDa protein--an in vivo comparison of physical enhancement methods*. Journal of drug targeting, 2010. **18**(2): p. 141-7.

# Chapter 17
# THE PRINCIPLES OF CELLULAR COMMUNICATION IN THE SKIN

Cellular communication is a rapidly developing discipline which, by its nature, is a highly complex and inaccessible subject to most people apart from biochemists and applied scientists. Therefore, only the principles of cellular communication and their practical applications in active skin care will be outlined here. Understanding these principles will make it much easier to understand the detail described in this book.

## COMMUNITIES OF CELLS

Skin cells, like humans, live in close contact with each other as individuals, as families, and in communities where the success of these communities is determined by the efficacy of their communication and how well they bind as physical individuals to achieve common goals.

All humans, wherever they are, develop from a single fertilized female egg cell. Fertilization sets a process in motion that consists of innumerable messages passing between cells to grow, develop, and maintain a large number of specialized and highly organized cellular communities with unique functions. The language on which cells depend is not yet clearly documented nor understood in the type of detail that would allow scientists and doctors to manipulate cell behaviour at will. However, in healthy and well-nourished skin the cells seem to thrive and 'understand' each other perfectly.

> *"Healthy thriving skin requires an endless communication of high-quality information."*

The cell communication 'language' is passed on from one generation of humans to the next through our genetic code. Each complete set of chromosomes is formed by a unique fusion of two genetic halves, one of which usually comes from outside the maternal family line. This mode of information transmission ensures that apart from minor individual variations, all skin cells require certain very specific molecules to maintain optimal communication structure and function. At the heart of this process is vitamin A, literally from conception to death.

Stem cells have to develop into very highly specialized cells, tissues, and organs that need to function harmoniously. Vitamin A acts as the coordinator of this process in concert with a great many other elements derived from adjacent cells, far distant endocrine organs, the immune system, the brain, and a variety of nutrients. The process of passing on the information, which orchestrates this complex collection of cell processes, does not change in successive generations of humans because of the way new human beings are conceived and the way genetic material containing the information is passed on. This process of stable, minimally-altering information transmission means that there exists a group of essential molecules that have become permanently indispensible to optimally active intercellular communications and the orderly functioning of the individual cells.

In skin the ones that may be labeled truly 'essential' and those that originate from nutrition are vitamins A, C, E, and D. They are 'essential' because without them, the rest of the signals involved (and there are many) will not function to preserve and promote cell development, specialization, and maintenance.

In skin care the absolute necessity of these core molecules is a vital element separating the active cosmetic formulations from the largely inactive ones. The essential molecules in skin are as crucial to cell communication as a keystone is in an arch on which the rest of the structure depends.

The second essential element is that communication between cells is based on a balanced mixture of a very large number of chemical 'words' and physical bonds. Chemical 'words' may be defined as groups of atoms or molecules that temporarily attach themselves to a variety of receptors on top of and inside cells to initiate specific, precise, actions in those cells. An analogous situation is a foreman using a variety of words strung together in understandable and logical sentences to instruct a diverse group of individuals to complete a complex task in construction, for instance.

On the cellular level the skin-cell receptors act as the sensory organs, just as ears and eyes do in humans, to take in the proffered instructions and respond appropriately. The interaction between signaling atoms or molecules, also called ligands and the receptors forms the basis of the larger part of cellular and intercellular communication. Direct contact between cells makes up the rest of it.

## COMMUNICATION AMONG CELLS

Healthy thriving skin requires an endless communication of high-quality information. Signaling between cells starts very early. For instance, for a fertilized ovum to implant successfully, a series of signals has to be exchanged between the developing embryo, called a blastocyst, and the lining of the uterus.[1] Without this communication, implantation of the blastocyst does not happen and the developing conceptus will be lost.

Signals exist in the form of atoms, molecules, and combinations of molecules and ensure that almost all of the component tissues that make up each human are kept highly specialized. In skin the important functions are protection, temperature regulation, disposal of some waste products, psychosocial functions, and environmental interaction. There are probably no quiet periods in skin-cell communication, but rather it is a continuous process from the first

### MODES OF CELL COMMUNICATION

❶ Cells communicate with each other via direct contact (juxtacrine signaling), over short distances (paracrine signaling), or over large distances and/or scales (endocrine signaling)[3]

❷ Some cell-to-cell communication requires direct contact between cells. Here cells can form gap junctions that connect their cytoplasm to the cytoplasm of adjacent cells, like passages through which the atoms and molecules serving as the words and sentences between cells may be freely exchanged.

❸ The notch signaling mechanism is an example of juxtacrine signaling (also known as contact-dependent signaling) in which two adjacent cells must make physical contact in order to communicate. This requirement for direct contact allows for very precise control of cell differentiation (specialization) during embryonic development.

❹ Many cell signals are conveyed by molecules that are released by one cell and move to make contact with another cell. Endocrine signals are called hormones. In the skin the gender hormones like oestrogen, progesterone, and testosterone help instruct cells over very large distances. Produced in the ovaries, testes, and adrenal glands these signals are carried to the skin cells by the blood. As they are hormones, these signals act directly on receptors situated on the DNA of cell nuclei. A number of molecules classed as inflammatory signals are part of the communication systems of the skin. They are produced by the immune system and no specific gland, but they nevertheless have important communication functions in the skin.

❺ The short distance active paracrine signals, such as retinoic acid, target only cells in the vicinity of the emitting cell.[5] As these signals need to be delivered in the vicinity of the cells, it makes sense to apply signals like vitamin A and C topically to the skin.

divisions of the fertilised egg to the death of the human.

## THE MEANING OF CELLULAR COMMUNICATIONS IN SKIN

### 'LANGUAGE USE'

Natural human languages contain collections of words which impart meaning by use of words individually, but more commonly as part of complex sentences containing rich meaning and often leading to actions of equal complexity and variety. In the human body the information exchange between its cells is no different, except that on the cellular level there is probably not much idle chatter.

It stands to reason that every cell or organ in the human body is intimately related to every other part, as all of the various human tissues merely exist as specializations of a single stem cell. It is conceivable, therefore, that seemingly unrelated cells communicate with each other actively to maintain the stable status quo of the entire organism. In other words, every single cell that makes up a human being, has a single, common ancestor. From this fact it follows that the therapist who wishes to restore skin to the highest level of function and specialization should look specifically for practical means to find elements missing in damaged or poorly functioning skin cells. The therapist can then restore or normalize the communication among cells by supplying the missing 'words

"In the human body the information exchange between its cells is no different, except that on the cellular level there is probably not much idle chatter."

and phrases' in the correct places and in the correct amounts, to achieve this lofty goal. Communications between cells forms the basis of active cosmetic treatment of skin.

Restoring malfunctioning or faulty skin cells is analogous to teaching victims of stroke to speak again. Instructing them to express themselves clearly will enable them to live as normally as possible in a speaking society. As long as the damaged brains miss key words in the language, their speech remains largely garbled and unintelligible, preventing their normalization of quality of life.

In skin the deficiency of key molecules like vitamin A, C, E, and D leads to skin which ages faster, does not heal efficiently, and becomes thin and emaciated. Pigment production and distribution becomes disturbed and the skin layers lose their unique, resilient, and sturdy binding to each other. Providing the deficient chemical 'words' in sufficient quantities restores the ability of the skin cells to reverse the negative trends mentioned above. Like speaking again after a stroke, restoration of age- and environment-damaged skin is a normalization process. If one has to use a single word to describe the core function that vitamin A performs then it is 'normalization'.

## PRINCIPLES OF SKIN TREATMENT THROUGH CELL COMMUNICATION

In focusing specifically on skin cell communication, therapists already have very specific and readily accessible resources to communicate with cells to achieve the normalization process in damaged skin. There are, however, important principles one needs to follow on a practical day-to-day level.

**I. THE FIRST PRINCIPLE** of paramount importance is never to attempt to use a single entity (a chemical 'word') in an attempt to restore a complex system usually reliant on a large number of cell signals for optimal function and structure.

A speech therapist cannot use the phrase 'I love you' as the only means of complete restoration of speech in a stroke victim. No doubt the phrase is an important one in human existence, but it has little relevance if a stroke victim keeps uttering only that one phrase all the time, in every social situation and in answer to every question put to him or her. Functional restoration will require a large, rich, and varied vocabulary, applied with correct syntax and in the appropriate context. Restoration of disturbed cell communication is no different. It requires a large, rich and varied 'chemical vocabulary', applied with correct syntax and in the appropriate context.

**II. THE SECOND PRINCIPLE** is the one that states: 'Where the body will do the job perfectly itself, one should not attempt to reinvent the wheel'. The practical implication of this principle for damaged or ageing skin is that one needs only to understand the deficiencies that may exist in such skin and the consequences of such shortcomings. There is simply no need to decipher and mimic the entire process. Here the second principle espouses the much greater potential and actual use of cell communication in practical day-to-day skin therapies.

The body has a proven ability to restore and replace damaged tissue, which is accessed through parts of a process known in biology as the 'wound healing cascade'. This elegant pathway represents cell communication in skin of the highest order. At least two easily-performed procedures can engineer this transition by activating the cascade in skin. These therapies lead one into the third and very powerful principle.

**III. THE THIRD PRINCIPLE** is that human biology retains the ability to repair tissue via stem-cell pathways from birth to death. This innate biological tissue restoration capacity of the skin is usually activated by injury. From the work of Whitby and Ferguson of Manchester in the UK we know, for instance, that it is entirely possible to let wounds heal by complete regeneration of tissue as opposed to healing with scarring.[2]

It is 'simply' a matter of understanding how to ensure the correct concentration of specific cell signals at specific times in tissue for the changes to come about. In other words, one has to ensure that the correct instructions are communicated to the cells responsible for replacing the injured tissue with new and normal tissue components.

'Foetal wound healing', as this process has been called, has a direct bearing on the use of cell communication

> "*Human biology retains the ability to repair tissue via stem-cell pathways from birth to death*"

in active cosmetic skin treatment, as this is a process from which invaluable information may be extracted to achieve better skin for a lifetime. Getting to grips with the complexities of precisely how to instruct the stem-cell lines (basal cells in the epidermis and fibroblasts in the dermis) is the practical hurdle scientists and doctors still have to clear.

**IV. PRACTICAL APPLICATIONS.** Presently 'percutaneous collagen induction' by needling and 'platelet-rich plasma' injections are two examples of advanced selective activation techniques of the 'wound healing cascade' by blood platelet induction widely practiced today. Both these treatments rely on a high level of cell communication to instruct the skin stem-cell lines to replace old and damaged components with new ones.

These processes do not run optimally when administered on their own, however. There will certainly be some effect if used in isolation, but it will probably be limited. These processes need to be 'fuelled' in the same way any efficient vehicle requires fuel to drive its engine.

Fortunately, through very good research into skin over the past 100 years, a number of things have been learned and have been put into efficient practical use over the past 30 years.

1. Enough Vitamin A in cell storage ensures a constant momentum for cells to specialize better.

2. Much of the damage in cells is caused by an excess of free-radical molecules or atoms, which can be countered by supplying 'free-radical mops' like vitamins C and E and a myriad of other antioxidants like resveratrol, beta-carotene and a large and varied group of molecules known as polyphenols.

3. Vitamin C is an indispensible cell instruction for maintenance of connective tissue and pigment formation.

4. Enough vitamin D procured through controlled sun exposure at the correct times ensures a healthy skeleton and healthy skin cells and helps prevent many cancers. When sunlight is not sufficient, an oral supplement will do.

5. Good general health, with balanced gender hormones, positive metabolic balance, and the absence of persistent inflammation of any kind in the body are of invaluable assistance to ensure effective and healthy cell communications.

## CONCLUSION

Using cell communication in the skin as a tool to assist skin to its best potential is no longer in the realm of science fiction. It is now science fact.

The use of recognized chemical 'words and phrases' to optimize the way cells communicate with each other has become commonplace in skin care, even if one has not been aware of the exact details.

Understanding and positively altering existing cell communications is a living, practical method of approach to optimal skin maintenance and repair.

Claims that one may individualize skin treatments not based on the clearly described and established systems of inter-cellular communication amount to little more than interpretative guesswork with no scientific explanation.

Communication between cells holds the key to future tissue and organ restoration. Describing the systems that are active in the skin may well provide much information for in vivo tissue engineering by 'speaking in the tongue of cells'. Active skin care need no longer dwell in the realms of whims and mystical botanical extracts. It is a logical and systematic application of knowledge based on fundamental and profound communication on the cellular level. The result of this metamorphosis is healthier, better quality, and certainly more beautiful skin for much longer.

## REFERENCES

1. Uterine Wnt-catenin signaling is required for implantation.
   Othman A. Mohamed*, Maud Jonnaert, Cassane Labelle-Dumais;, Kazuki Kuroda;Hugh J. Clarke* and Daniel Dufort; Departments of Biology and Obstetrics and Gynecology and Division of Experimental Medicine, Royal Victoria Hospital, McGill University, Montreal, QC, Canada H3A 1A1; and Ottawa Hospital Research Institute, Ottawa, ON, Canada K1Y 4E9
2. Immunohistochemical studies of extracellular matrix growth factors in fetal and adult wound healing' In "Fetal Wound Healing: A Paradigm for Tissue Repair Whitby DJ & Ferguson MWJ (eds. N.S. Adzick & M.T. Longaker), Elsevier Science Publishing Co., New York, 161-175, 1991
3. http://en.wikipedia.org/wiki/Cell_signalling
4. Minimally Invasive Percutaneous Collagen Induction – Desmond Fernandes, MB,BCh, FRCS(Edin) Oral Maxillofacial Surg Clin N Am 17 (2005) 51 – 63 ; abme.com.br/pdfs/Dermaroller/maxilofacial.pdf
5. Duester G (September 2008). "Retinoic Acid Synthesis and Signaling during Early Organogenesis". Cell 134 (6): 921–31.2008.

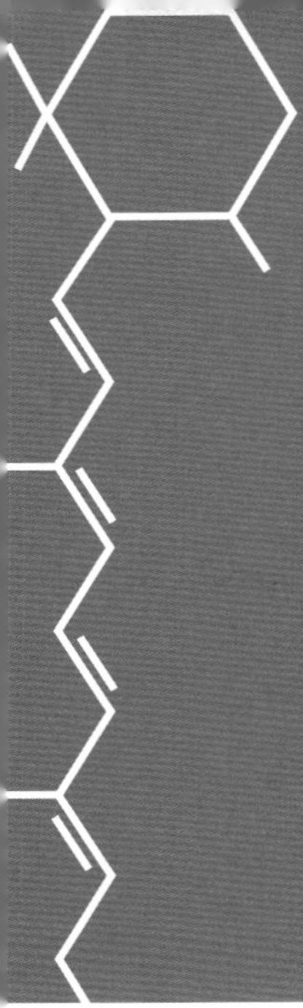

# Chapter 18
# SKIN NEEDLING

**SUMMARY:**

Needling skin has become an established method to rejuvenate skin and refine scars. While the technique may seem new, we actually have centuries of experience of needling skin because tattooing has been practiced in both 'civilized' and 'primitive' cultures all around the world.

Needling stands above all other current treatments because it has been shown to be the first method described, as far as we can determine, that conclusively regenerates tissue and restores the natural collagen lattice of the dermis. To date it is the only skin treatment that regenerates elastin. It has a wide list of indications and can be used on all ages and all different coloured skins with safety. For results one has to cause bleeding of the skin because it is the release of platelets that stimulates the regeneration of tissue, and the degree of improvement is directly related to the amount of bleeding. Des Fernandes has been needling himself and his patients since 1996, and has learned some pearls and pitfalls that he will share in this chapter.

## INTRODUCTION

Fernandes started experimenting with needling of the skin in the early 1990s and by 1996 started the micro-needling of skin in the way that we now do it. The story of skin needling is basically the report of his work, so this chapter will be more biographical than usual. Fernandes soon noticed that we were getting changes that could only be interpreted as being due to the release of growth factors from platelets, as in the normal wounding of skin. While at the time there was no evidence to confirm these deductions, fortunately over time research has substantiated these ideas.[1] The reason that he came to this conclusion was that several years earlier he had wondered why treatment of skin with retinyl palmitate (the predominant form of vitamin A in the body) and vitamin C could apparently restore collagen and thicken skin. The induction of growth factors seemed the logical answer. Time has also proved that vitamin A, on its own, stimulates impressive thickening of skin and increased collagen deposition.[2] Right from the beginning he used topical vitamin A and C creams (made by Environ Skin Care) on all the patients that he needled. They used the creams for about three months before the needling treatment and then continued to use them afterwards without stopping.

## THE MECHANISM FOR RELEASE OF GROWTH FACTORS

A single needle prick through the skin would cause an invisible response. A completely different picture emerges when thousands of fine pricks are placed next to each other. The bleeding automatically initiates the release of platelets and a complex chemical cascade of platelet-derived growth

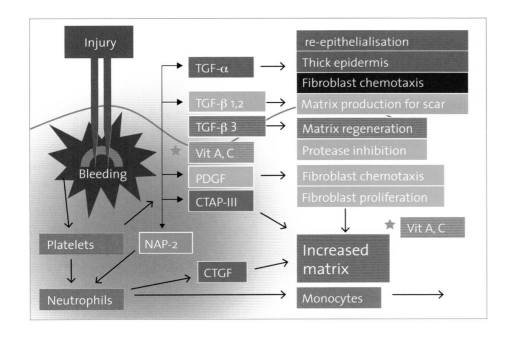

*Bleeding initiates the release of platelets that have the primary action of staunching the bleeding. Platelets at the same time release a raft of growth factors that basically hasten healing and the development of a strong matrix of fibres.*

> "Today Fernandes has the longest follow-up in the world and the number of cases is continually growing."

factors (PDGF) that set up a chain reaction with the eventual production of increased dermal-matrix proteins and a thicker epidermis. Particularly important is TGF-beta3, which is selectively released in preference to TGF-beta1 and 2 because this is a closed wound. TGF-beta3 has been shown by Ferguson and his team to be the responsible agent in scarless healing.[3, 4]

Fibroblasts migrate into the area, and this surge of activity inevitably leads to the production of more collagen and more elastin. Keratinocytes migrate rapidly across the minute epidermal defect and then proliferate, so the epidermis becomes thicker. In ordinary open wounds, collagen III is mainly produced, and research from Aust confirms that predominantly collagen I in lattice formation, which is normally found in young people, is laid down.[5] This is normally seen only with regenerating foetal tissue, where there is no evidence of any scar tissue when the baby is born. This is the landmark difference between skin-needling and other ablative or minimally-ablative procedures. Needling definitely causes regeneration rather than scar formation.[6]

The first presentation on needling of skin raised a lot of interest in 1999 at the International Plastic, Reconstructive and Aesthetic Society (IPRAS) in San Francisco. Fernandes called the process 'collagen induction therapy' and that term is now used around the world and usually abbreviated to CIT. Subsequently textbooks and other articles have been written about CIT.[7-11] Some people have complained that this is very autobiographical, but the fact is that no one else in the world was doing skin needling in this way and there was no other prior research to turn to.

Since 1996 CIT has grown from a treatment that was considered barbaric and medieval to a procedure that is now performed everywhere. Today Fernandes has the longest follow-up in the world and the number of cases is continually growing. In the process, experience has shown how to employ needling in a better way. This experience is important because when needling is done by inexperienced people the results will not be reliable, and in order to preserve the reputation of skin needling it has to be understood and done properly. The more minds that are applied to the concept of CIT, the more we will learn about how to do it better.

## THERE ARE THREE TYPES OF NEEDLING OF SKIN

When Fernandes started needling he first used a tattooing type of device that allowed him to experiment with varying depths of penetration. He soon realized that very superficial needling (0.5 mm) did not deliver noticeable results. It was less painful and hardly bled, and that is probably why there was not an adequate response. Simple electrical changes to membranes do not seem to stimulate the system as well as expected. He then realized that the final result was directly proportional to the degree of bleeding; the more bleeding holes, the better the result.

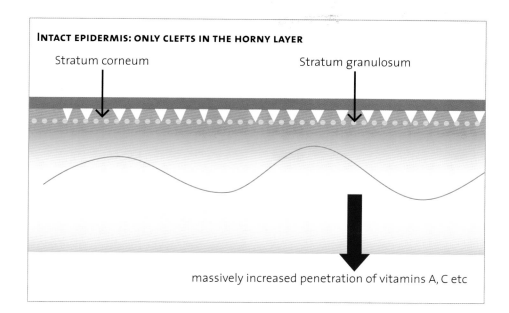

*Diagram of the skin, showing triangular wedges representing needle holes that only penetrate the stratum corneum. This is the best microneedling for enhanced penetration.*

**1. 'COSMETIC' SKIN-NEEDLING:** here we use needles that are preferably 0.1 or at maximum 0.2 mm long to penetrate only the stratum corneum and upper stratum spinosum at the deepest. At this depth the needling can barely be felt. A sufficient number of holes in the stratum corneum, the only barrier to the penetration of topically applied products, can achieve a massive enhancement of penetration into the lower levels of the skin. The more holes the better the penetration; the depth is less important. The only barrier to penetration lies in the stratum corneum and once the needle gets into the upper layer of the stratum spinosum, the skin is 'opened', and it is very easy for a molecule to move right through into the dermis. Because the number of needle holes determines the degree of enhanced penetration, you should needle as intensively and as frequently as possible in order the allow the active molecules to make changes in the deeper cells of the skin. Using a 0.1 mm device means this can easily and safely done every single day.

You can induce collagen formation only if the ingredients of a topically applied skin product can do it. Vitamin A in all its forms induces fibroblast DNA to produce the proteins that end up in normal lattice-type collagen I. Vitamin C has some effect in promoting collagen-inducing genes but is also essential as a co-factor in the production of normal collagen type I and III. A few peptides will also induce collagen production by their cytokine-like effects on keratinocytes and fibroblasts.

If you do not apply a product containing these ingredients, you

*Two years needling plus vitamin A and C and Matrixyl. Step 3 is after two years.*

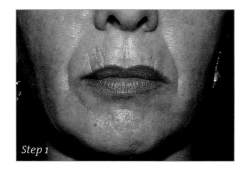
Step 1

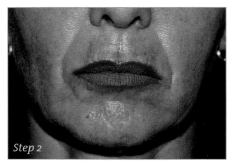
Step 2

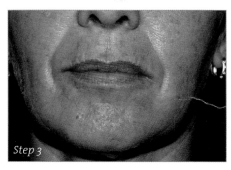
Step 3

cannot expect any result, except that the skin will be pricked, causing a slight bit of swelling that will puff out fine wrinkles temporarily. One has to look at long-term results to assess the clinical results scientifically.

## SIDES OF EYES

The cosmetic rollers have been complemented by the cosmetic stamping devices, which are smaller and devoted to treating smaller areas. At this stage there are no equivalent copies of this device from other companies. The needles of the Focus-CIT all penetrate vertically into the skin. The Focus-CIT is useful to enhance penetration both of specialized products to treat pigmentation problems and of other ingredients that cannot normally penetrate skin, like hyaluronic acid or muscle-relaxing molecules. Cosmetic needling is best called 'home mesotherapy' because the clients or patients themselves do the enhancement of penetration, and the

*Tightening of the skin at the lateral side of the eyes has occurred in a 65-year old woman as result of using cosmetic needling to enhance penetration of her vitamin A and antioxidant cosmetic over a span of one year.*

Before

After

intensity of their treatment determines the result. For best effects it should be done at least once a day. Of course, the subject should be barely aware of the needle pricks making holes. If the needles are too long, the subject will not needle adequately and the result will be impaired. When needles penetrate only 0.1 mm, they are barely perceptible and patients usually needle more thoroughly.

This is the first roller used for CIT since 1999, and it replaced the simple tattoo device. The device was made for Environ Skin Care and has subsequently been copied by many companies. The principle remains the same whatever device is used and provided the best fine stainless steel needles are used, the results achieved are not dependent on the manufacturer. Of course there are varying degrees of quality. Caveat emptor.

**II. THE MEDICAL ROLL/ FOCUS-CIT** has needles that protrude 1.0 to 1.5 mm so that the needles will penetrate right down into the dermis and rupture the tiny arcade capillaries in the dermal papillae.

While the Cosmetic Roll-CIT is not painful, Medical Roll-CIT is because the needles reach the cutaneous nerves. For that reason we have to use topical anaesthesia to permit adequate but light needling of the skin for about five to ten minutes in each area. Fernandes has shown that by alkalinizing the skin at the time of cleaning it, any topically applied anaesthetic will work more efficiently. The reason for this is that products containing anaesthetic agents are formulated at an acid pH

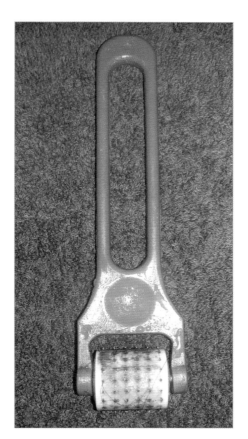

*1 mm roller after needling treatment.*

*"No roller device can be patented because of a prior patent in the 1950s by Pistor, the 'father' of mesotherapy."*

to keep the anaesthetic agent stable. However, anaesthetic agents work best in an alkaline medium. Fernandes has formulated special products to alkalinize the area to be needled prior to applying the anaesthetic product.

If we want to use intensive Medical Roll-CIT then in some people we should use regional anaesthesia or nerve blocks. In that case one can treat thoroughly and make as many holes as possible in the dermal blood vessels. If one uses a mechanical tattooing type of device, even topical anaesthesia seems to last

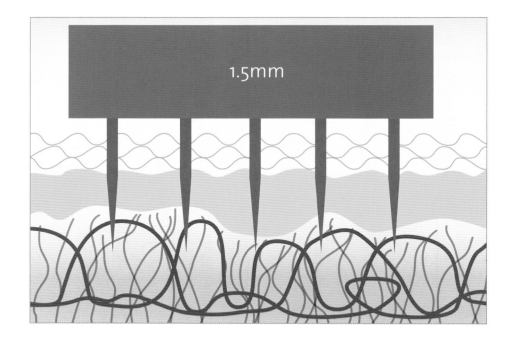

*Diagram of Medical Focus CIT that penetrates through the epidermis and punctures blood vessels in the papillary dermis*

longer, but then one has to beware of doing too heavy a treatment that will interfere with the patient's normal social activities. The more one needles, and the more one bleeds, the better the result. In 1996 Fernandes initially used a tattoo artist's device (a simpler version of more sophisticated electrical needling devices) for intensive 1 to 1.5 mm needling, but that proved to be too arduous and also risky. This technique can excoriate skin, but fortunately tattoo artists are well trained to avoid this. Because Fernandes recognized the difficulties of mechanical devices, he designed a rolling device that has now been copied in many countries worldwide. In addition Fernandes felt that the needles had to penetrate deeper and thereby fracture significantly more blood vessels. In order to penetrate with a number of needles at the same time, they have to be spaced so that the skin resistance does not inhibit penetration. Fakirs in India lie on a bed of nails, but the nails do not penetrate the skin because there are so many sharp nails so close together. If the nails were farther apart, they would penetrate the skin.

No roller device can be patented because of a prior patent in the 1950s by Pistor, the 'father' of mesotherapy. Any patents claimed are only weak ones relating to needle anchoring and the like. That is why so many companies have copied the device. However, whatever company's device is used, the results of needling will be the same if the needles penetrate into the dermis.

**III. THE SURGICAL ROLL-CIT** has needles that penetrate 3 mm, and

Before

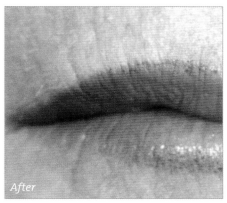

After

*View of the right side of the upper lip in detail before and one year after one medical needling. Not only are the creases on the upper lip reduced but the volume of the lip has also become plumper without any filler.*

people are generally quite frightened by its appearance. Treatments using it have to be done with full anaesthesia of the skin. This device is used for intensive treatments and is essential for treating deep, thick acne scars or burn scars. It is also very useful for treating the abdomen, breasts, and limbs, because one can apply less pressure compared to the shorter needle devices to get adequate needling. There is a body device with two angled rollers (also in 0.1, 1.5, and 3 mm sizes) for treating curved surfaces. The abdomen, breasts, and limbs often get sufficient anaesthesia from topical anaesthetic products, so a general anaesthetic is not necessary unless extensive areas or the whole body is being done. Skin-needling can be used for total body rejuvenation.

**IV. 3 MM ROLLERS ARE MOST USEFUL FOR BURN SCAR TREATMENTS.** The needles are long enough to penetrate through the thick scars and induce bleeding at deeper levels.

A very detailed 'road-map' for skin needling can be found in the *"Illustrated Guide to Percutaneous Collagen*

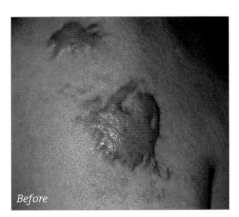

Before

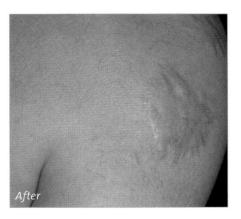

After

*Hypertrophic burn scars respond well to skin needling. This demonstrates a typical response to 3 mm needling of a 2-year-old flame burn scar on the shoulder. Two 3 mm needling treatments were done and the after picture was taken about six months after needling. He used vitamin A,C and E oil daily.*

Skin Needling / 195

> "3 mm rollers are most useful for burn scar treatments. The needles are long enough to penetrate through the thick scars and induce bleeding at deeper levels."

*Induction: Basics/ Indications/ Uses* by Matthias Aust, Svenje Baht and Des Fernandes. Publication Date: *August 15, 2013* | ISBN-10: 1850972532 | ISBN-13: 978-1850972532 | 1st Edition:

Every practitioner in skin needling should use this book.

The skin is routinely prepared by using topical vitamins A and C and antioxidants for at least three weeks, but it is better for patients to prepare their skin for three months to get the best results, and always if the skin is very sun-damaged. Vitamin-A-based products must have adequate levels of vitamin A and C. Effective levels of the vitamins may be indicated by a tendency to cause a retinoid reaction. If the stratum corneum is thickened and rough, a series of mild TCA peels will prepare the surface of the skin for needling and maximize the result. Fernandes used a unique cream peel product with 2.5 per cent to 5.0 per cent TCA.

Under topical, local, or general anaesthesia, the skin is closely punctured with the specific rolling or mechanical device, which consists of a rolling barrel with needles at regular intervals. For deeper needling one should avoid a dense array of needles so that skin resistance will not prevent the penetration. By rolling backwards and forwards in various directions one can achieve an even distribution of the holes. Alternatively, use the Focus-CIT for smaller areas. Since the needles penetrate the epidermis but do not remove it, the epidermis is only punctured and will rapidly heal. The skin bleeds for a short while, but that soon stops. Of course, the skin develops multiple micro bruises in the dermis. It is essential to use wet gauze swabs to soak up any ooze of serum especially when 3 mm needling has been done. This serous ooze can be stopped faster by using a TCA 2.5 per cent cream for four minutes to get effects without creating a deeper peel. Once the serous ooze has stopped, the skin is washed thoroughly and then covered with a special vitamin A, C, and E oil (do not use retinoic or ascorbic acid!). Patients are warned that they will look terribly red and bruised if the needling has been done intensively with a 3-mm device. If they have had a 1-mm needling, they will look sunburned but bruising is uncommon. If the mechanical tattoo-style device is used even at 1.5 mm or 2 mm, there is more swelling than with the 1.5 mm roller and more downtime. The patients who have had a lighter 1-mm needling can return to work almost immediately after the treatment. Patients are encouraged to shower within an hour of the procedure when they return home. A recommended gel cleanser with tea tree oil has been found, by experience with many hundreds of patients, to be very useful and safe.

## CARE OF THE SKIN AFTER PERCUTANEOUS CIT

Immediately after a 3-mm needling the skin looks bruised, but bleeding is minimal and there is only a small ooze of serum that soon stops. After all, this is only a pin-prick! The patient is encouraged to use topical vitamin A (retinyl ester) and vitamin C ester cream immediately to take advantage of the increased penetration and to promote better healing and greater production of collagen. While there was no proof in the beginning, subsequent research by M. Aust has confirmed that needling alone was almost 50 per cent less effective than needling with vitamin A skin care.[6] The addition of a special cocktail of peptides has ensured even better results.

With 3-mm needling, the skin feels tight and might look uncomfortable, but it is not. The next day the skin looks less dramatic, and by day four to five the skin has returned to a moderate pink flush that can be concealed with make-up. Iontophoresis and sonophoresis of vitamin A and C immediately after the needling treatment maximizes the induction of healthy collagen. Iontophoresis also tends to reduce the swelling of the skin. Low-frequency sonophoresis can be used to enhance penetration of peptides to induce even more collagen and elastin production.

This is a simple technique, and with the right tool it is easy and fast to puncture any skin thoroughly. While one

### THE INDICATIONS FOR COLLAGEN INDUCTION THERAPY ARE:

To correct skin laxity and restore tightness in the early stages of ageing. This is a relatively minor procedure and can safely be recommended. Some patients who are reluctant to do facelift surgery may be satisfied with simple percutaneous CIT. The arms, abdomen, thighs, and buttocks can also be treated. In fact, this has become the only anti-ageing full-body treatment that we have.

1. To repair fine wrinkles from photoageing.
2. To treat acne scarring. The skin becomes thicker and the results are superior to dermabrasion, which makes the skin thinner.
3. To remove or diminish stretch marks, which respond well to skin needling even when they have become 'silvery' with time.
4. To tighten skin after liposuction. This is a major innovation that ensures a more youthful surface as well as the better shape from liposuction.
5. To improve scars, even if they are old white scars they can become more skin-coloured. Even burn scars, both flat and hypertrophic, can be treated with success. In 2010 Matthias Aust was awarded the European Burns Association Prize of 10,000 euros for his work on burn scars.
6. To treat skin-grafted insensitive scars; after a few needlings it seems as though the cutaneous nerves are restored, because the skin becomes much more sensitive. Presumably one of the growth factors released is responsible for nerve regrowth.

### ADVANTAGES OF CIT

1. CIT does not damage the skin. The epidermis remains intact.
2. Any part of the body may be treated.
3. The skin becomes thicker, as has been repeatedly demonstrated in histological studies.
4. The healing phase is short.
5. It is not as expensive as laser resurfacing.
6. The skin does not become sun-sensitive.
7. It can be done on people who have had laser resurfacing or those with very thin skin.
8. Hyperpigmentation in patients with darker skins like African, Indian, Malaysian, Chinese, and Mediterranean skins has not been seen, though some patient worry about transient hyperpigmentation. In general, pigmented blemishes become lighter coloured.
9. Telangiectasia may completely disappear.
10. It does not have to be done by a doctor but should be done under medical supervision.
11. The technique is easy to master with rollers that have been specially designed for the procedure. The Focus-CIT type of device is useful for smaller areas.
12. It can even be done with topical anaesthesia.

treatment may not give the smoothing seen with ablative laser resurfacing, the epidermis is normalized, and if the result is not sufficient it can be repeated as often as necessary. The technique can be used on areas that are not suitable for peeling or laser resurfacing.

### SOME IMPORTANT PEARLS

1. Always prepare the skin with a reliable vitamin-A-based product and use it both day and night. The best products use retinyl esters, which are photo-protective and absorb UV rays. The most effective product in my experience with effects comparable to using retinoic acid is Environ Skin Care products. This is the product that has been used in the research of Fernandes and Aust.

2. Prepare the skin for as long as possible before the needling. The longer the patient has been on vitamin A, the better and faster it will heal. The skin also becomes more tolerant of injury. The higher the dose of vitamin A the better; however, always ensure that the doses gradually increase so that they can stimulate the production of more cellular retinoid receptors and thereby avoid retinoid skin reactions and produce better results.

3. There must be bleeding from the tiny vessels in the vascular arcades

## DISADVANTAGES OF CIT

1. If the procedure is done properly there will be exposure to blood, so the proper precautions have to be followed.
2. We cannot achieve as intense a deposition of collagen as in laser resurfacing (but then again, we can repeat the treatment and get even better results that will last just as long, if not longer).
3. Over-aggressive needling with a tattoo-artist's style mechanical device may cause scarring. Too close and dense a treatment will excoriate the epidermis.
4. Local anaesthetics limit the area that may be treated, and in some cases a general anaesthetic is required, as, for example, when extensive areas or virtually the whole body is needled.

of the dermal papillae. It is not enough to just see the skin become flushed; you need evidence of bleeding, and the best evidence is some degree of bleeding from the surface. Some people insist that the growth factors are not derived from platelets but are released instead as a result of cellular electrical changes from puncturing the cells. While growth factors might indeed be released, the chances are that the quantity is relatively insignificant. Experience over time has shown that the less one bleeds, the less impressive is the result.

4. If the patient has acne spots at the time of needling as a result of pre-operative stress, then treat the skin with a mild acid such as Environ ACM 2 for ten minutes prior to the procedure. This will cause small areas of 'frosting' at the infected spots that effectively sterilize the skin.

5. At the end of the needling session add a mild peeling treatment using very weak doses of acid as found in the Environ ACM 1 (which is half the strength of 2) for five minutes to ensure that the patient does not get any infected spots in the treated area. If acne-prone skin is not treated this way, pimples may develop, whereas when using this added light peel, pimples never occur.

## SOME IDEAS TO MAXIMIZE THE RESULT OF PERCUTANEOUS CIT

1. By experimenting with the various time schedules, Fernandes found that increasing the frequency of needling treatments from once a year to once a month eventually to once a week produced greater tightening and smoothing of skin. Studies by Zeitter et al have shown that when animals were needled at weekly intervals, the level of growth factors increased beyond expectation as compared to the levels in the similar animals after

*Ink was applied to the surface after 1 mm needling and allowed to dry. We can see extensive, passive penetration and diffusion of the ink right into the dermis.*

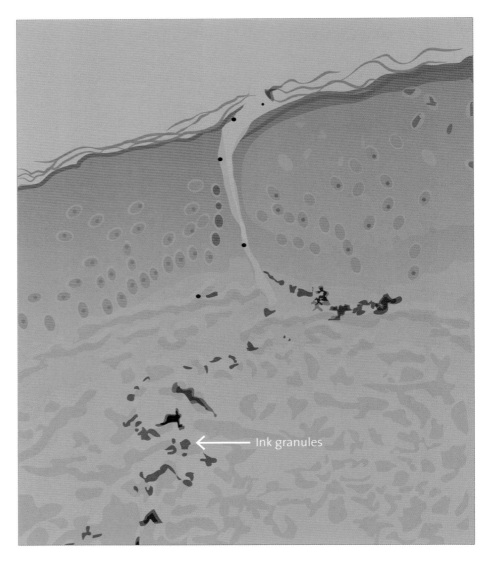

Ink granules

one needling.[12] This finding opens up many possibilities in treating people with topical anaesthesia in a shorter period. It is important to recognize that especially when one does needling at frequent intervals it is utterly essential to ensure that the skin is supported with topical vitamin A as well as vitamin C. In Zeitter's study they recommended Environ vitamin A, C and E Oil and C-Boost.

One of the problems about skin needling is that too few people have an extended and large experience of needling. To date, Aust in Germany[5,6,11,13-16] and El Domyati in Egypt seem to have the only reasonable series in clinical and laboratory studies of skin needling, but they are in animals. Non-invasive studies by Fernandes et al. are in the process of publication testing even shorter intervals in between needling.

❷ Immediately after a needling session is an excellent time to apply a topical hyaluronic acid product. Hyaluronic acid, which cannot penetrate the stratum corneum, would normally lie on the surface, whereas after needling the hyaluronic acid easily penetrates into all layers of the skin. Studies done by Fernandes have shown that anything placed on the surface of the skin after needling easily penetrates 1 to 2 mm into the dermis. (See diagram) A medium-chain hyaluronic acid can stimulate the CD 44 receptor on the keratinocytes and facilitate the production of more GAGs in the skin.[17] Patients find hyaluronic acid extremely soothing immediately after needling.

❸ Immediately after the needling session, patients should ideally have sonophoresis with iontophoresis of selected peptides to further enhance collagen and elastin production. Fernandes has worked with peptide complexes and in presentations has shown rapid and more intensely successful outcomes for skin needling at 1 to 3 mm.

❹ After the peptide treatment or instead of peptides, Fernandes recommends the patient should have iontophoresis of vitamin A and C to support intense collagen and elastin production.

❺ Intense pulsed light (IPL) immediately after needling has given longer lasting and effective results according to half-face studies. The IPL is used immediately after the treatment session to take advantage of the haemoglobin chromophores lying in the dermis. The relevance of this in regard to regeneration has not been explained. In a similar way, a half-face study (unpublished) using LED therapy with 633 and 830 nm got approval from 80 per cent of the patients, who felt that the LED-treated side had a better result. The device used was Omnilux ReviveTM (633nm) therapy that according to research, stimulates fibroblast activity, leading to faster and more efficient collagen synthesis and ECM proteins and also increases cell vitality by increasing the production of cellular ATP. Omnilux plusTM (830nm) stimulates the contractile phase of the remodelling process producing better lineated collagen. In combination these wavelengths work synergistically to produce better visible results and hence patient satisfaction.

❻ Platelet-derived growth factor (PDGF) may offer another way of enhancing the release of growth factors and thereby achieving a better result. At this stage no one has published any conclusive results. Fernandes uses PDGF in conjunction with skin needling as often as possible in the belief that it does contribute to a better result.

## SOME PITFALLS TO AVOID

❶ Never use retinoic acid after needling. It irritates the skin and the patient is also less likely to use it properly. Several individuals around the world who have used retinoic acid with skin needling have reported that they had difficulties and were disappointed with the result.

❷ Never use ascorbic acid preparations immediately after needling. Fernandes produced a very deep peel that caused a depigmented area on the patient's face when he applied ascorbic acid to the surface immediately after needling.

❸ Never use strong peeling agents after needling. The peel will be very much deeper than expected and may cause scars.

❹ Needling is painful for about 20 minutes after the operation, so if the procedure has been done with general anaesthesia and without local anaesthetic infiltration, do make sure that the patient is given adequate analgesia before waking up.

## CONCLUSION

Skin needling induces collagen and elastin production by employing the body's natural mechanisms that for the first time, as far as we know, produces regeneration of the skin and its matrix. There is no scar formation and the procedure can safely be repeated until the desired effect is achieved.

## REFERENCES

1. Aust, M.C., et al., *Percutaneous collagen induction-regeneration in place of cicatrisation?* J Plast Reconstr Aesthet Surg, 2010.
2. Yan, J., et al., *Levels of retinyl palmitate and retinol in the skin of SKH-1 mice topically treated with retinyl palmitate and concomitant exposure to simulated solar light for thirteen weeks.* Toxicol Ind Health, 2007. **23**(10): p. 581-9.
3. Shah, M., D.M. Foreman, and M.W. Ferguson, *Neutralising antibody to TGF-beta 1,2 reduces cutaneous scarring in adult rodents.* Journal of cell science, 1994. **107**(Pt 5): p. 1137-57.
4. Durani, P., et al., *Avotermin: a novel antiscarring agent.* Int J Low Extrem Wounds, 2008. **7**(3): p. 160-8.
5. Aust MC, R.K., Vogt PM, *Medical needling: improving the appearance of hypertrophic burn scars.* GMS Verbrennungsmedizin, 2009. **3**(Doc 03).
6. Aust, M.C., et al., *Percutaneous collagen induction-regeneration in place of cicatrisation?* Journal of plastic, reconstructive & aesthetic surgery : JPRAS, 2011. **64**(1): p. 97-107.
7. Fernandes, D. and M. Signorini, *Combating photoaging with percutaneous collagen induction.* Clin Dermatol, 2008. **26**(2): p. 192-9.
8. Fernandes, D., *Percutaneous collagen induction: an alternative to laser resurfacing.* Aesthet Surg J, 2002. **22**(3): p. 307-9.
9. Fernandes, D., *Minimally invasive percutaneous collagen induction.* Oral Maxillofac Surg Clin North Am, 2005. **17**(1): p. 51-63.

10. Aust, M.C., et al., *Percutaneous collagen induction therapy: an alternative treatment for scars, wrinkles, and skin laxity.* Plast Reconstr Surg, 2008. **121**(4): p. 1421-9.

11. Aust, M., et al., *Percutaneous collagen induction therapy for hand rejuvenation.* Plastic and reconstructive surgery, 2010. **126**(4): p. 203e-204e.

12. Zeitter, S., et al., *Microneedling: Matching the results of medical needling and repetitive treatments to maximize potential for skin regeneration.* Burns : journal of the International Society for Burn Injuries, 2014. **40**(5): p. 966-73.

13. Aust, M.C., et al., *Percutaneous collagen induction. Scarless skin rejuvenation: fact or fiction?* Clinical and experimental dermatology, 2010. **35**(4): p. 437-9.

14. Aust, M.C., et al., *Percutaneous collagen induction therapy: an alternative treatment for burn scars.* Burns : journal of the International Society for Burn Injuries, 2010. **36**(6): p. 836-43.

15. Aust, M.C., K. Knobloch, and P.M. Vogt, *Percutaneous collagen induction therapy as a novel therapeutic option for Striae distensae.* Plastic and reconstructive surgery, 2010. **126**(4): p. 219e-220e.

16. Aust, M.C., et al., *Percutaneous collagen induction: minimally invasive skin rejuvenation without risk of hyperpigmentation-fact or fiction?* Plastic and reconstructive surgery, 2008. **122**(5): p. 1553-63.

17. Barnes, L., et al., *Synergistic effect of hyaluronate fragments in retinaldehyde-induced skin hyperplasia which is a Cd44-dependent phenomenon.* PloS one, 2010. **5**(12): p. e14372.

# Chapter 19
# A FINAL WORD DR FERNANDES

I hope that by reading this book you will have gained insight into the way skin care should be applied to produce beautiful skin for a lifetime.

In the world out there we will continually be confronted by so called 'new and revolutionary' treatments. It is important to know that most of these treatments will come and go, because they will usually not be based on the science of how the skin really works. It is only by studying skin cells and structures carefully in the laboratory and clinic that we can understand how to improve the way skin elements function best.

There is a lot more to discover and refine in the quest for healthier and more beautiful skin. However, the fundamental elements we have discovered, analyzed, and learned to address will not change, as the way they are inherited from one generation to the next is genetic.

I have no doubt that as we learn more about the 'language of cells' our ability to formulate more powerful and precise instructions will improve greatly. Nor do I have any doubt that vitamin A will always be central to any advances that are made.

There are interesting and important differences between, for example, Asians, Caucasians, Africans, etc., but these differences are actually quite subtle. They challenge us merely to work out the best treatments, but make no difference to the way in which skin treatment needs to be approached. Respecting the structures of skin and the harmony that exists in the integrity of the various elements that make up this beautiful laminated structure, is the key to the philosophy we should follow when repairing damaged skin or when preserving good skin.

It is never too early in life to start thinking about skin care. From the day of birth, skin is the canvas of life that presents our picture to the world. Taking the best care of this valuable organ improves quality of life on all levels. Our parents and grandparents did not have the knowledge to educate our generation about the best way to keep skin healthy and beautiful, but we have that knowledge now, and we have developed the tools to teach our children to have much better skin in generations to come. This information places the responsibility on all of us to educate children about the science and practice of vitamin-A-based skin treatment.

In closing, I would like to say that I am filled with gratitude by the knowledge that literally thousands of people have benefited from the work I have invested in this skin-care approach over the last 30 years. As young doctors we often hope that we will succeed in making the lives of other people better in one way or another. Seeing any person's skin changing from a damaged or aged state to a healthy, comfortable and beautiful complexion is an experience that will always give me a sense of pleasure and profound satisfaction.

I am privileged to have this road to walk.

**Des Fernandes**
**Cape Town**
**(February 2015)**

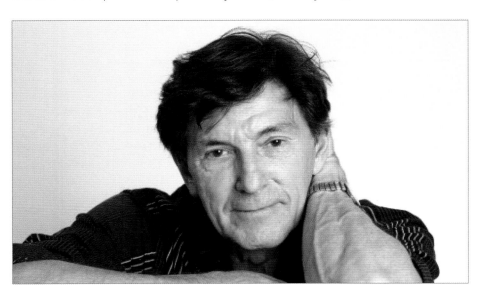

# Vitamin A
# SKIN SCIENCE INDEX

## A

Ablative laser   198
Ablative procedures   190
Abnormal pigmentation   98, 113, 116, 145, 156, 157, 171
Abrasions   144, 168
Abuse of animals   161
Abdomen, needling   194, 195, 197
Acetyl hexapeptide-3 (Argireline®)   134, 135
Acetylcholine   134, 135
Acetic acid   140
Acetylsalicylic acid   146
Acidity of skin   144
ACM 1   199
ACM 2   199
Acne   18, 23, 67, 76, 85, 86, 89, 92, **93,** 94, 111, 112, 138, 141, 144, 146, 147, 152, 153, 154, 156, 160, 166, 168, 169, 170, 195, 197, 199
Acne scarring   160, 195, 197
Actinic keratoses   47, 67, 102, 155
Adrenal glands   33, 183
Advanced Glycation end products (AGEs)   50
Advanced penetration   159, 171
African skin   29, 198
Ageing   15, 16, 17, 20, 23, 30, 32, 36, 39, 42, 43 45, 48, 55, 61, 67, 71, 75, 81, 86 95, 118, 129, 143, 147, 168, 185, 197
Ageing cells   168
AGE's   50
Ageing skin   16, 45, 81, 143, 185
Age spots   95, 96
Aggressive peel   151
AHA   18, 51, 97, 98, **138-148**, 151

AHA use guidelines  145
Alanine  127
Alberts D.  102
Alcohol  18, 65, 84, 103, 105, 111, 118
Alginate mask (masque)  165, 170, 174
Alkalinizing skin for topical anaesthesia  193
Allergy sensitization  51
All-trans 3, 4-Dihydroretinoic acid  69
All-trans retinoic acid  69
Alpha hydroxy acid (**AHA**)  18, 51, 92, 93, 109, **138-148**
Alpha-lipoic acid  42, 43, 60, 113, 119, 109, **118**
Alpha tocopherol  112
Alumina  59
Amazonian Indians  135
Amino acid  **127-129**, 132
Ammonium lactate (neutralized lactic acid)  98, 145, 146, 176
Anion  110
Antiageing  104, 122, 197
Anti-carcinogenic effects  116, 124, 125
Antioxidants  12, 18, 32, 40, 41, 42, 43, 51, 55, 56, 57, 59, 60, 61, 81, 83, 91, 92, 93, 94, 95, 96, 97, 104, **106-119**, 122, 123, 124, 125, 128, 129, 130, 141, 143, 145, 153, 154, 155, 156, 160, 169, 186, 192, 196
Antioxidant brigade  41, 42, 107, 111, 123
Antioxidant network  42, 43, 117, **118**
Antioxidant recycling  **118**
Apocrine sweat glands  33
Aqueous channels  173
Arcade capillaries  193
Argireline (peptide)  130, 131, 132, 133, **134**, 135
Ascorbic acid  18, 43, **106-112**, 131, 141, 160, 166, 168, 177, 196, 202
Ascorbyl di-palmitate  111
Ascorbyl salts  109

Ascorbyl tetra-isopalmitate  110, 111, 112
Asian skin  27, 29, 116, 117, 204
Aspalanthus linearis leaf extract (Rooibos tea)  123
Aspirin  146, 147
Atomic chlorine  166
Atopy  12, 94, 95
Aust, Mathias  190, 196, 197, 198, 201
Australian Tea Tree Oil  94, 196

## B

Barrier  25, 26, 35, 36, 47, 68, 82, 123, 129, 130, 132, 142, 144, 146, 159, 191
Barrier function of skin  159
Basal cell carcinomas  89
Basal cells  48, 102, 186
Basal layer  **24**, 25, 26, 73, 82, 102, 141, 148
Basal membrane  25, 31
Basement membrane  24, 38, 135, 136
Bemotrizinol (Tinosorb-S)  58
Benzophenone-3  57, 58
Benzophenone-34  57
Beta-carotene  12, 13, 20, 43, 58, 60, 66, **71**, 84, 85, 90, 91, 94, 117, 122, 174, 178, 186
Beta-hydroxy acid  18, 94, 138, 139, **145-148**
Biotin  124, 125
Birth defects  76, 103
Bisoctrizole (Tinasorb-M)  58
Blackheads  33, 48, 85, 86, 92, 93, 144, 147
Black skin  27, 28, 84
Blue light  18, 36, 49, 51, 117
Body needling device  195
Botanical extracts  187
Botulinum-like affect  130
Breasts  needling  195
Broccoli  71, 118
Burn  28, 29, 39, 60, 91, 166, 168, 169, 170, 175, 195, 197
Burn scar  91, 195, 196, 197

## C

Caffeine   123
Calcium   116, 117, 123, 134, 135
Callused skin   145
Cancer   11, 13, 25, 27, 28, 35, 36, 37, 38, 39, 42, 43, 47, 48, 50, 54, 55, 56, 57, 59, 61, 74, 80, 81, 82, 84, 86, 96, 101, 102, 103, 105, 112, 116, 117, 118, 119, 122, 123, 124, 128, 151, 156, 186
Cancer   prevention 56
Cancer chemoprevention   102, 103
Cancer prone cells   151
Carboxylic acid   139, 141
Carnosic acid   125
Carotenoids   42, 60, 66, 71, 73, 84, **117**
Carotenoid timeline   **66-67**
Caucasian skin   28, 29, 47
Cavitation   171, **172**, 173, 174, 175, 176, 177
Cavitation bubbles   172, 174, 176
Cavitation rate   **175**
CD44 receptor   30, 31, 201
Cell communication   181, 182, 183, 185, 186, 187
Cell differentiation   8, 183
Cell senescence   81
Cell signal   98, 147, 183, 185
Cell specialization   186,
Cell turnover   147, 148
Cellular debris   148
Cellular communication   **181-187**
Cellulite   136
Ceramide   82, 142
Chemical barrier of skin   132
Chemical penetration enhancement   131
Chiral molecule   112
Chloasma   95
Chromophore   49, 50, 115, 201
Chromosomes   25, 32, 84, 123, 182
Chickenpox scars   108

CIT (collagen induction therapy)   31, 132, 190, 192, 193, 194, **196-199**
Citric acid   138, 146, 147
Chicken skin   47
Chinese skin   198
Chronoageing   129
Cis-retinoic acid   49, 69, 76, 94
Cis-urocanic acid   50
Clefts in the stratum corneum   191
Cluver   58, 67, 71, 81, 83
Cod-liver oil   66
Coenzyme Q10   42, 43, 60, 113, **117-118**
Co-factor   110, 115, 191
Communities of cells   181
Conjunctiva   85
Connective tissue   186
Collagen   18, 20, 24, 29, **30-32**, 34, 35, 42, 46, 47, 50, 59, 60, 74, 75, 82, 85, 92, 93, 107, 108, 110, 111, 122, 124, 129, 130, 134, 135, 141, 146, 147, 151, 152, 153, 163, 168, 186, 188, 189, 190, 191, 195, 197, 199, 201, 202
Collagenase   29, 50, 124, 125, 134
Collagen degeneration   60
Collagen formation   59, 191
Collagen inducing genes   191
Collagen induction   186, 191, 197
Collagen induction therapy   190, 197
Collagen lattice   188
Collagen synthesis   147, 201
Collagen type I   191
Collagen type III   191
Colostrum   178
Comedones   48
Congenital malformations   103
Contact-dependent signaling   183
Continuous galvanic current   168, 169, 170
Convection (of molecules)   176
Cool peeling   **151**, 152, 153, 154, 155

Copper   123, 130, 134
Copper tripeptide   130, 134
Corneocytes   48, 172
Cortisone   13, 29, 33, 94, 95
Cosmeceutics   141
Cosmetic skin needling (0.1 -0.2 mm)
    34, 130, 157, **191**
Cosmetic Roll CIT®   31, 132, 193
Cosmetic stamping device   192
Couperose   144, 169
Coupling medium in LFS cavitation   174
Curare   135
Cutaneous nerves   193, 197
Cyanocobalamin   115
Cyclopia genistoides leaf extract
    (Organic Honey Bush tea extract)   123
Cystine   127
Cytokines   128

## D

D-ascorbic acid   107
Deep phenol peel   150, 151
7-Dehydrocholesterol   50
Dehydro-ascorbic acid   108
Dermabrasion   197
Dermal layers   86, 146, 147
Dermal matrix proteins   190
Dermal papillae   22, 25, 34, 74, 82, 136,
    150, 193, 199
Dermis   21, 22, 23, 24, 25, 29, 30, 32, 33, 34,
    35, 37, 38, 39, 46, 49, 50, 51, 72, 74, 85,
    86, 104, 107, 119, 135, 136, 141, 146, 148,
    150, 151, 156, 160, 166, 170, 186, 188, 191,
    193, 194, 196, 200, 201
Dermal blood vessels   193
Desmosome   25, 140, 145
Dibenzoylmethane (Parsol 1789)   58
Dietary supplements   61
Differentiation   69
Dihydrotestosterone   33
Disencrustation   167
Disposal of waste products   183
Disturbed cell communication   185
D-lactic acid   139
DMSO (penetrant enhancer)   176
DNA   42, 48, 50-51, 57, 73, 74, 80-81, 85,
    86, 107, 110, 112, 113, 114, 123, 124, 134,
    141, 183, 191
DNA Mutations   112
DNA Oxidation   61
DNA repair   113
DNA synthesis   114
De-differentiation of cells   13
D-panthenol   114
Dipeptide   128, 129
Dipeptide diaminobutyroyl (Sny-ake
    peptide)   135
Dry skin   144

## E

Ebers papyrus   64
Eczema   12, 94,
Egyptian papyrus   150,
Elastin   20, 31, 42, 50, 75, 107, 110, 122, 129,
    130, 134, 136, 152, 153, 190, 197, 201, 202
Elastases   50
Elastosis   47, 60, 92, 93
Electrical burn   169
Electro-osmosis   165, 167
Electroporation   165
Eleidin   26
Emulsifiers   131
Endocrine signaling   183
Energy metabolism   113
Enhanced skin penetration   157, **158-178**,
    191
Environ Ionzyme DF machine   174
Environ Ionzyme DF II machine   175
Enzymes in skin for vitamin A   76
Epidermal atrophy   60

Epidermis   21, **24**, 25, 34, 50, 73-75, 86, 108, 119, 129, 136, 146, 147, 150, 151, 152, 153, 158, 160, 170, 173, 186, 194, 196, 198
Epidermal cells   144
Epigallocatechin gallate (EGCG)   124
Ethoxy-ethanol   131
Eumelanin   26
Excess free radicals   186
Exogenous ochronosis   156, 157

## F

Fergusson   190
Fibroblast   **29**, 74-75, 81, 107, 124, 129, 141, 186, 190, 191
Fibronectin   29, 32
Field iontophoresis   170
Fitzpatrick skin colour scale   **28**, 37
Flavonoids   60
Focus CIT®   132, 192, 194, 196, 198
Formation of collagen & elastin   31
14-hydroxy-4, 14-retroretinol   69
Free radical(s)   10, 18, 20, 27, 39, **40**, 41, 42, 50-51, 57, 58 , 82, 107, 113, 117 118, 119, 156, 186
Frosting   199
Fulton   67

## G

GAGs   25, 29, 30, 75, 82, 129, 146, 152, 201
Galvanic current   160, 163, 166, 170
Gamma-tocopherol   112
Gap junctions   183
Glutamic acid   128
Glutathione   43, 49, 60, 109, 113, 114, **118**, 124, 128
Glycine   127, 128
Glycation   50, 147
Glycation end products   50
Glycolic acid   51, 138, 139, 141, 142, 146, 147, 148, 151, 166

Glycosaminoglycans (**GAGs**)   25, 29, 30, 75, 82, 141, 146, 168
Granular layer   **25**
Green tea Extract   123, 124
Growth factors   122, 128-130, 152, 174, 178, 189, 197, 199
Growth hormone   128

## H

Hair follicles   25 148, 173
Hayflick-limit   74
Heavy peeling   156
Herpes simplex   168
Home peels   94
Homocystein   114, 115
Honey   64
Hormonal pigmentation (melasma)   52, 80, 83, 94, 96, 110
Horny layer   **26**, 35, 54, 82, 85, 98, 109, 140, 141, 144, 146, 147, 158, 173
Hyaluronic acid   **30**, 35, 75, 168, 201
Hyaluronan   **30**, 35
Hydroquinone   157
Hyperkeratotic skin   142
Hypertrophic burn scar   195
Hypertrophic scar   167, 197
Hypo-vitaminosis A   80, 103

## I

Idebenone   118
Infra-red light damage   38, 59, 117
Intense pulsed light (IPL)   201
Intercellular communication   182
Intermittent galvanic current   164
Inorganic reflectant sunscreens   56, 58, 59, 156
Ionisation   160
Iontophoresis   69, 108, 110, 111, 132-133, 157, **158-170**, 197, 201

## J
Juxtacrine signaling   183

## K
Karyometric image analysis   102
Keratin   32, 33, 82, 85
Keratinocytes   24, 46, 50, 74, 81-83, 96, 113, 124, 129, 141, 150, 190, 191
Keratin plugs   147
Keratolytic action   146
Kligman Prof A   67, 83, 146

## L
Lactic acid   98, 138, **142**, 145, 155, 166, 170
Lacunae (small lakes)   173
L-ascorbic acid   165, 167
L-lactic acid   139
Lamellar bodies   172
Laminin   32
Langer's lines   22,
Langerhans cells   25, 34, 39, 47, 57, 81, 82, 124
Lattice (collagen)   74, 107
Lentigo   83, 95
Leuphasil (peptide)   130, 132, **134**
LFS (Low Frequency Sound)   166, **175-178**
LFS complications   **175**
LFS machines   **173**
Ligand   182
Light peeling   153
Linoleic acid   176
Lipid bilayers   144, 146, 172, 176
Lipid-peroxidation   50, 124
Lipofuscin   95
Liposomes   68, 131
Local anaesthesia   196, 199
Local anaesthetic cream   174
Low-concentration acid peels   152
Low frequency sonophoresis   197
L-panthenol   114

Lupus erythematosis   116, 168
Lutein   60, 117
Lycopene   58, 60, 117
Lymph vessels in skin   **34**

## M
Magnesium (or sodium) ascorbyl phosphate   109, 110
Maleic acid   138
Malignant melanoma   8, 10, 11, 43, 80, 89, 124, 130
Mangiferin   123
Matrix-metallo endoprotease (**MMPs**)   50, 74, 82
Matrixyl   130, **134**, 168, 192
Medical focus CIT (1 to 1.5mm)   193
Medical Roll CIT (1 to 1.5mm)   193
Melanin   26, 29, 49, 50, 54-55, 74, 92, 93, 95, 96, 108, 110, 113, 130, 141, 157
Melanocortin   26
Melanocytes   26, 27, **29**, 51, 60, 80, 81, 93, 95, 113, 157
Melanophores   27, 34
Melanosomes   27, 34
Melasma   52, 80, 83, 94, 96, 97, 110
Messenger chemicals   128
Methoxycinnamate   57
Methylcobalamin (Vit B12)   114
Microdermabrasion   159
Micro-needling   31, 159, 160, 170, 188, 191
Minimal erythema dose   37
Mini peel   170
Mitragotri S. (sonophoresis)   133, 171
Mitochondria   41, 114, 117, 118
Mitochondrial DNA   118
MMPs   50, 74, 82
Muscle-relaxing molecules   192

## N
NADH   49

Nails   33, 125
Natural Acid mantle   143
Natural sunscreen   29, 50, 80, 92, 113, 115, 116, 124
Nature-identical   139, 142
Needling   **188-202**
Negative charge   160
Neuromuscular junction   134
Neuro-peptide   175
Neurotransmitter-affecting peptides   130
Niacinamide   176
Nicotinamide   49, 113
Nicotinamide adenine dinucleotide (NAD)   113
Night blindness   64, 65, 104
9-Cis-retinoic acid   69
Notch signaling   183

## O

Ochronosis   156, 157
Octocrylene   57
Oligo-peptides   127
Oily skin   93, 147
Organic honeybush tea extract   123
Organic sun filters   **56–61**
Organic sunscreen   156
Ornithine decarboxylase   74
Ozone   36, 54
Oxygen   40, 41, 118, 119

## P

PABA   57
Packer, Lester   118
Palmitoyl pentapeptide (Matrixyl)   134
Palmitoyl oligopeptide (Vialox®)   135
Panthenol (Vit B5)   114
Pantothenic acid   114
Paracelsus   100
Paracrine signaling   183
Parsol-1789   58

Papillary dermis   21, 194
Para-amino benzoic acid (PABA)   57
Peeling for acne   153, 154
Peeling of skin   68, 94, 110, 154, 198
Pellagra   113
Penetrant enhancer   75, 131, 132 -133
Penta-peptide-3 (Leuphasil®)   **134-135**
Percutaneous collagen induction   186, 197
Pernicious anaemia   114
Peptide(s)   20, **127-137**, 168, 178, 191, 197
Peptide naming   128
Phaeomelanin   26
Phenol peeling   150
Phenol toxicity   151
Phonophoresis   171
Plastic wrap   171
Platelet-derived growth factor   189, 202
Platelets   188, 189, 199
Platelet-rich plasma injection   186
Polypeptides   127, 128
Polyphenols   186
Proline   107
Propylene glycol   176
Pro-vitamin B5   91, 94
Poison   100, 162
Pre-vitamin D   37, 50
Protein kinase-C   113
Polypodium   58
Psoriasis   116
Pyrimidine dimers   124
Pyruvic acid   138, 139

## Q

## R

Rabbit ear   162
RAR vitamin A nuclear receptors   69
Ratio of vitamin A isomers   80
Reflectors   **56-61**, 96

Reflectant minerals   56
Regenerative mechanism   152, 188
Regeneration of skin   202
Reiss   66
Repetitive serial light peeling   157
Resveratrol   122, 186
Rete pegs and ridges   25, 136
Retina   65
Retinal (Retinaldehyde)   69, 84
Retinol   **65**, 67-77, 81, 167, 168, 177, 178
Retinoic acid   17, 18, 65, 66, 67-77, **69-75**, 81, 83, 89, 105, 134, 143, 146 167, 168, 178, 202
Retinoic acid receptor alpha   103
Retinoic acid receptor beta   103
Retinoic acid receptor X   103
Retinoids   65, 71, 76, 80, 82, 136
Retinoid reaction   18, 76, 91, 132, 145, 196, 198
Retinoid receptors   68, 76, 82, 91, 102, 103, 198
Retinyl acetate   68, 69 -70, **72-75**, 160
Retinyl aldehyde   **69**, 71-73, 81
Retinyl esters   70-74, 83, 197, 198
Retinyl palmitate   20, 49, 50, 51, 58, 61, 64, 65, 68-77, **69-75**, 80, 97, 131, 154, 160, 167, 189
Retinyl palmitate oil   154
Retinyl propionate   69
Riboflavin   49, 50
Rosacea   144
Rosemary extract (carnosic acid)   125
Rooibos tea   96, 123
RXR vitamin A nuclear receptors   69

# S
Safety of vitamin A   100-105
Salicylic acid   94, 146, 147, 148, 151
Salicylic acid toxicity   146
Salicylism   147

Scar collagen   151
Scurvy   106
Sebaceous activity   153
Sebaceous follicle   33, 92, 141
Sebaceous glands   22, 33, 144, 147
Sebum   16, 85, 94, 147, 167
Selenium   60, **119**
Serum levels from topical vitamin-A   76
7-Dehydrocholesterol   50
Signaling molecules   20, 130, 182
Silica   59
Singlet oxygen   41, 124
Skin cancer   13, 42, 43, 48, 50, 54, 56-57, 74, 75, 80, 82, 96, 105, 151, 155, 156
Skin cancer chemoprevention   102, 103, 124
Skin colour   **26**
Skin layers   21, 34
Skin needling   **188-202**
Skin occlusion   174
Skin peeling   **150-157**
SLES   176
SNAP complex   134,
Sodium ascorbyl phosphate   110, 167, 168
Sodium channels   135
Sodium lauryl sulphate (SLS)   174
Solar Comedones   48
Solar keratoses   42, 43, 47, 89, 96, 156
SPF   29, 46, 56
Squamous cell cancer   102
Stem cells   13, 24, 69, 73, 80, 184
Stem cell repair pathways   185
Stratum corneum   **26**, 73, 82, 110, 128, 129, 132-**133**, 142, 143, 146, 148, 164, 165, 166, 172, 173, 176, 190, 191, 196, 201
Stratum lucidum   26
Stratum spinosum   **25**, 83, 124, 144, 191
Stratum granulosum   **25**, 191
Stretch marks   197
Stuttgen   67

Sub-cutaneous fat layer   21 20
Surgical Roll CIT (3mm)   194
Sulzberger   80
Sunburn cells   113
Sun exposure   48, 45, **54-61**, 76, 147
Sun freckles   46, 56
Sunscreen   43, **55-61**, 92, 93, 156
Sunspots   47, 82
Superficial needling (0.5mm)   190
Superoxide dismutase   60, **119**, 123
Sweat gland ducts   150, 173
Sweat glands   22, 33, 150
Syn-Ake (peptide)   130, 132, 135
Synthetic vitamin E   112

## T

TCA Peel(s)   196
Telomere shortening   48, 112
Telangiectasia   47, 198
Telomerase   81
Telomeres   81
TGF-beta1   190
TGF-beta2   190
TGF-beta3   190
Therapeutic ultrasound   171
Threshold energy density in LFS   174
Tinasorb-M   58
Tinosorb-S   58
Titanium dioxide   56, 59
Titanium oxide   59
TNF-$\alpha$   147
Tocotrienol   113
Tomato   43, 58, 71, 125, 138
Transforming growth factor beta-1   190
Transforming growth factor beta-2   190
Transforming growth factor beta-3   74, 190
Trans-resveratrol   122
Trans-retinoic acid   49
Tumour necrosis factor $\alpha$ (TNF-$\alpha$)   147

Tyrosinase   26, 96, 108, 122, 141, 142, 167
Tyrosinase inhibitor   108, 142, 167

## U

Ultrasound   133, 159, 170
UVA B and C   36, 38, **39**, 49, 54, 72, 83, 115, 116, 117, 124, 125, 156
UV Damage   37 58, 75, 80, 82-83, 113, 122, 124
UV damage repair   75
UVA and B penetration   55-56, 39, 72
UVA protection   111
Urocanic acid   50
UV Light   29, 36, 43, 54-62, 76, 90, 107, 198
UV Damage   **54-61**, 125

## V

Vascular arcades   198
Violet light   52
Violox (peptide)   130, 132, **135**
Visual purple   65
Vitamin A metabolism   **80**
Vitamin A paradox   159
Vitamin B3, (Nicotinamide)   113
Vitamin B5 (Panthenol)   **114**
Vitamin B12 absorption through mouth   114
Vitamin B12 (Methylcobalamin)   **114**, 174, 178,
Vitamin C   18, 43, 51, 58, 60, 90, 93, 94, **106-112**, 118, 123, 125, 141, 153, 186, 189, 191, 197
Vitamin C deficiency   107
Vitamin C ester   197
Vitamin C in pigmentation   108
Vitamin D   17, 50, 55, 69, 80, 83, **115-117**, 182, 186
Vitamin D and UVA   37, **116**
Vitamin D deficiency   38, 48, 55, 61, 116
Vitamin D toxicity   117

Vitamin E   43, 50, 58, 60, 90, 94, **112-113**, 118, 123, 125
Vitamin E reactivation   113
Vitamin K   103

## W

Waterproofing effect in skin   34, 35, 82, 141
Water-repellent   110
White heads   92, 93, 147

## X

Xanthone   123,
Xerophthalmia (dry eyes)   66, 85, 103

## Y

## Z

Zinc   61, **119**, 123
Zinc oxide   56, 59